Biochemistry
FOR
DUMMIES®
2ND EDITION

by John T. Moore, EdD, and Richard Langley, PhD

WILEY

Wiley Publishing, Inc.

Biochemistry For Dummies®, 2nd Edition

Published by
Wiley Publishing, Inc.
111 River St.
Hoboken, NJ 07030-5774
www.wiley.com

WILEY

About the Authors

John Moore grew up in the foothills of western North Carolina. He attended the University of North Carolina at Asheville, where he received his bachelor's degree in chemistry. He earned his master's degree in chemistry from Furman University in Greenville, South Carolina. After a stint in the U.S. Army, he decided to try his hand at teaching. In 1971 he joined the chemistry faculty of Stephen F. Austin State University in Nacogdoches, Texas, where he still teaches chemistry. In 1985 he started back to school part time, and in 1991 he received his doctorate in education from Texas A&M University. He has been the co-editor (along with one of his former students) of the "Chemistry for Kids" feature of *The Journal of Chemical Education.* In 2003 his first book, *Chemistry For Dummies* (Wiley), was published, soon to be followed by *Chemistry Made Simple* (Broadway Books). John enjoys cooking and making custom knife handles from exotic woods.

Richard Langley grew up in southwestern Ohio. He attended Miami University in Oxford, Ohio, where he received bachelor's degrees in chemistry and mineralogy and then a master's degree in chemistry. His next stop was the University of Nebraska, where he received his doctorate in chemistry. Afterward, he took a postdoctoral position at Arizona State University in Tempe, Arizona, followed by a visiting assistant professor position at the University of Wisconsin at River Falls. In 1982, he moved to Stephen F. Austin State University. For the past several years, he and John Moore have been graders for the free response portion of the AP Chemistry Exam. He and John have collaborated on several writing projects, including *5 Steps to a 5 on the AP: Chemistry* (McGraw-Hill), *Chemistry for the Utterly Confused* (McGraw-Hill), and *Organic Chemistry II For Dummies* (Wiley). Rich enjoys jewelry making and science fiction.

Dedication

To my wife, Robin; sons, Matthew and Jason; my wonderful daughter-in-law, Sara; and the two most wonderful grandkids in the world, Zane and Sadie. I love you guys. — John

To my mother. — Rich

Authors' Acknowledgments

We would not have had the opportunity to write this book without the encouragement of our agent, Grace Freedson. We would also like to thank Vicki Adang for her support and assistance on this project. Thanks to our colleague Michele Harris, who helped with suggestions and ideas. Thanks to Britney Cooper, who helped us with proofreading. We're also grateful to our technical editors, Mary Peek and Sara O'Brien, for their comments and contributions. And many thanks to all the people at Wiley Publishing, who helped bring this project from concept to publication.

Publisher's Acknowledgments

We're proud of this book; please send us your comments at http://dummies.custhelp.com. For other comments, please contact our Customer Care Department within the U.S. at 877-762-2974, outside the U.S. at 317-572-3993, or fax 317-572-4002.

Some of the people who helped bring this book to market include the following:

Acquisitions, Editorial, and Media Development

Project Editor: Victoria M. Adang
(Previous Edition: Kristin DeMint and Corbin Collins)

Acquisitions Editor: Stacy Kennedy

Copy Editor: Todd Lothery
(Previous Edition: Josh Dials and Corbin Collins)

Assistant Editor: David Lutton

Editorial Program Coordinator: Joe Niesen

Technical Editors: Sara O'Brien, PhD; Mary Peek, PhD

Editorial Manager: Michelle Hacker

Editorial Assistants: Rachelle Amick, Alexa Koschier

Art Coordinator: Alicia B. South

Cover Photo: ©iStockphoto.com/Martin McCarthy

Cartoons: Rich Tennant
(www.the5thwave.com)

Composition Services

Project Coordinator: Patrick Redmond

Layout and Graphics: Carl Byers, Carrie A. Cesavice, Nikki Gately, Corrie Socolovitch, Christin Swinford

Proofreaders: Laura L. Bownan, Jessica Kramer, Shannon Ramsey

Indexer: Sharon Shock

Publishing and Editorial for Consumer Dummies

    Diane Graves Steele, Vice President and Publisher

    Kristin Ferguson-Wagstaffe, Product Development Director

    Ensley Eikenburg, Associate Publisher, Travel

    Kelly Regan, Editorial Director, Travel

Publishing for Technology Dummies

    Andy Cummings, Vice President and Publisher

Composition Services

    Debbie Stailey, Director of Composition Services

Contents at a Glance

Table of Contents

Introduction

· ·

*W*elcome to the second edition of *Biochemistry For Dummies!* We're certainly happy you've decided to delve into the fascinating world of biochemistry. Biochemistry is a complex area of chemistry, but understanding biochemistry isn't really complex. It takes hard work, attention to detail, and the desire to know and to imagine. Biochemistry, like any area of chemistry, isn't a spectator sport. You must interact with the material, try different explanations, and ask yourself why things happen the way they do.

Work hard and you'll get through your biochem course. More important, you may grow to appreciate the symphony of chemical reactions that take place within a living organism, whether it's a one-celled organism, a tree, or a person. Just as each individual instrument contributes to an orchestra, each chemical reaction in an organism is necessary, and sometimes its part is quite complex. However, when you combine all the instruments and each instrument functions well, the result can be a wonder to behold. If one or two instruments are a little out of tune or aren't played well, the orchestra still functions, but things are a little off. The sound isn't quite as beautiful or there's a nagging sensation of something being wrong. The same is true of an organism. If all the reactions occur correctly at the right time, the organism functions well. If a reaction or a few reactions are off in some way, the organism may not function nearly as well. Genetic diseases, electrolyte imbalance, and other problems may cause the organism to falter. And what happens then? Biochemistry is often the field in which ways of restoring the organism to health are found and cures for many modern medical maladies are sought.

About This Book

Biochemistry For Dummies is an overview of the material covered in a typical college-level biochemistry course. In this second edition we attempted to update the material and correct the errors and omissions that crept into the first edition. We hope that this edition is of even more help than the first. We've made every attempt to keep the material as current as possible, but the field is changing ever so quickly. The basics, however, stay the same, and that's where we concentrate our efforts. We also include information on some of the applications of biochemistry that you read about in your everyday life, such as forensics, cloning, gene therapy, genetic testing, and genetically modified foods.

As you flip through this book, you see a lot of chemical structures and reactions. Much of biochemistry revolves around knowing the structures of the molecules involved in biochemical reactions. Function follows form. If you're in a biochemistry course, you've probably had at least one semester of organic chemistry. You'll recognize many of the structures, or at least the functional groups, from your study of organic chem. You'll see many of those mechanisms that you loved (and hated) here in biochemistry.

If you're taking a biochemistry course, use this rather inexpensive book to supplement that very expensive biochemistry textbook. If you bought this book to gain general knowledge about a fascinating subject, try not to get bogged down in the details. Skim the chapters. If you find a topic that interests you, stop and dive in. Have fun learning something new.

Conventions Used in This Book

We organize this text in a logical progression of topics that may be used in a biochemistry course. Along the way, we use the following conventions to make the presentation of information consistent and easy to understand:

- ✔ New terms appear in *italic* and are closely followed by their definition.
- ✔ We use **bold text** to highlight keywords in bulleted lists.

We also make extensive use of structures and reactions. While reading, try to follow along with the associated figures.

What You're Not to Read

Don't read what you don't need. Concentrate on the area(s) in which you need help. If you're interested in real-world applications of biochemistry, by all means read those sections (indicated by the Real World icon). However, if you just need help on straight biochemistry, feel free to skip the applications.

We also include some interesting topics in sidebars, the shaded boxes you find in many chapters. In those, you get a more in-depth look at some nonessential areas of biochem.

You don't have a whole lot of money invested in this book, so don't feel obligated to read everything. When you're done, you can put it on your bookshelf alongside *Chemistry For Dummies, The Doctor Who Error Finder,* and *A Brief History of Time* as a conversation piece.

Foolish Assumptions

We assume — and we all know about the perils of assumptions — that you're one of the following:

- ✔ A student taking a college-level biochemistry course
- ✔ A student reviewing your biochemistry for some type of standardized exam (the MCAT, for example)
- ✔ An individual who wants to know something about biochemistry
- ✔ A person who's been watching way too many forensic TV shows

If you fall into a different category, we hope you enjoy this book anyway.

How This Book Is Organized

Here's a very brief overview of the topics we cover in the various parts of this book. Use these descriptions and the table of contents to map out your strategy of study.

Part I: Setting the Stage: Basic Biochemistry Concepts

This part deals with basic aspects of chemistry and biochemistry. In the first chapter you find out about the field of biochemistry and its relationship to other fields within chemistry and biology. You also get a lot of info about the different types of cells and their parts. In Chapter 2 we review some aspects of water chemistry that have direct applications to the field of biochemistry, including pH and buffers. Finally, you end up with a one-chapter review of organic chemistry, from functional groups to isomers.

Part II: The Meat of Biochemistry: Proteins

In this part we concentrate on proteins. We introduce you to amino acids, the building blocks of proteins. Having the building blocks in hand, in the next chapter we show you the basics of amino acid sequencing and the different types of protein structure. We finish this part with a discussion of enzyme kinetics, both catalysts (which speed up reactions) and inhibitors (which slow them down).

Part III: Carbohydrates, Lipids, Nucleic Acids, and More

In this part we show you a number of biochemical species. You'll see that carbohydrates are far more complex than that doughnut you just ate may lead you to believe, but we do show you some biochemistry that is just as sweet! Then we jump over to lipids and steroids. Next are nucleic acids and the genetic code of life with DNA and RNA. Then it's on to vitamins (they're involved more than once a day) and hormones (no humor here — it would just be too easy).

Part IV: Bioenergetics and Pathways

It all comes down to energy, one way or another. In these chapters we look at energy requirements and where that energy goes. This is where you meet our friend ATP and battle the formidable citric acid cycle. Finally, because you'll be hot and sweaty anyway, we throw you into the really smelly bog of nitrogen chemistry.

Part V: Genetics: Why We Are What We Are

In this part we tell you all about making more DNA, the processes of replication, and several of the applications related to DNA sequencing. Then it's off to RNA and protein synthesis.

Part VI: The Part of Tens

In this final part of the book we discuss ten great applications of biochemistry to the everyday world and reveal ten not-so-typical biochemical careers.

Icons Used in This Book

If you've ever read a *For Dummies* book (such as the wonderful *Chemistry For Dummies*), you'll recognize most of the icons used in this book, but here are their meanings anyway:

The Real World icon points out information that has a direct application in the everyday world. These paragraphs may also help you understand the bigger picture of how and why biochemical mechanisms are in place.

This icon is a flag for those really important points that you shouldn't forget as you go deeper into the world of biochemistry.

We use this icon to alert you to a tip on the easiest or quickest way to learn a concept. Between the two of us, we have almost 70 years of teaching experience. We've learned a few tricks along the way and we don't mind sharing.

The Warning icon points to a procedure or potential outcome that can be dangerous. We call it our Don't-Try-This-At-Home icon.

Where to Go from Here

The answer to where you should start really depends on your prior knowledge and goals. As with all *For Dummies* books, this one attempts to make all the chapters discrete so that you can pick a chapter containing material you're having difficulty with and get after it, without having to have read other chapters first. If you feel comfortable with the topics covered in general and organic chemistry, feel free to skip Part I. If you want a general overview of biochemistry, skim the remainder of the book. Dive deeper into the gene pool when you find a topic that interests you.

And for all of you, no matter who you are or why you're reading this book, we hope that you have fun reading it and that it helps you increase your understanding of biochemistry.

Part I

Setting the Stage: Basic Biochemistry Concepts

"I love this time of year when the biochem students start exploring new and exciting ways for bonding carbon atoms."

In this part . . .

We go over some basic aspects of chemistry, organic chemistry, and biochemistry. First we survey the field of biochemistry and its relationship to other disciplines within chemistry and biology. We cover several different types of cells and their parts. Then we look at some features of water chemistry that apply to biochemistry, paying attention to pH and buffers. In the end, you get a brush-up on your organic chemistry, which sets the stage for Part II.

Chapter 1

Biochemistry: What You Need to Know and Why

*I*f you're enrolled in a biochemistry course, you may want to skip this chapter and go right to the specific chapter(s) in which we discuss the material you're having trouble with. But if you're *thinking* about taking a course in biochemistry or just want to explore an area that you know little about, keep reading. This chapter gives you basic information about cell types and cell parts, which are extremely important in biochemistry.

Sometimes you can get lost in the technical stuff and forget about the big picture. This chapter sets the stage for the details.

Why Biochemistry?

We suppose the flippant answer to the question "Why biochemistry?" is "Why not?" or "Because it's required."

That first response isn't a bad answer, actually. Look around. See all the living or once living things around you? The processes that allow them to grow, multiply, age, and die are all biochemical in nature. Sometimes we sit back and marvel at the complexity of life, fascinated by the myriad chemical reactions that are taking place right now within our own bodies and the ways in which these biochemical reactions work together so we can sit and contemplate them.

When John learned about the minor structural difference between starch and cellulose, he remembers thinking, "Just that little difference in the one linkage between those units is basically the difference between a potato and a tree." That fact made him want to learn more, to delve into the complexity of the chemistry of living things, to try to understand. We encourage you to step back from the details occasionally and marvel at the complexity and beauty of life.

What Is Biochemistry and Where Does It Take Place?

Biochemistry is the chemistry of living organisms. Biochemists study the chemical reactions that occur at the molecular level of organisms. Biochemistry is normally listed as a separate field of chemistry. However, in some schools it's part of biology and in others it's separate from both chemistry and biology.

Biochemistry really combines aspects of all the fields of chemistry. Because carbon is the element of life, *organic chemistry* plays a large part in biochemistry. Many times biochemists study how fast reactions occur — that's an example of *physical chemistry*. Often metals are incorporated into biochemical structures (such as iron in hemoglobin) — that's *inorganic chemistry*. Biochemists use sophisticated instrumentation to determine amounts and structures — that's *analytical chemistry*. And biochemistry is also similar to *molecular biology;* both fields study living systems at the molecular level, but biochemists concentrate on the chemical reactions that occur.

Biochemists may study individual electron transport within the cell, or they may study the processes involved in digestion. If it's alive, biochemists study it.

Types of Living Cells

All living organisms contain cells. A *cell* is not unlike a prison cell. The working apparatus of the cell is imprisoned within the "bars" — known as the *cell membrane.* Just as a prison inmate can still communicate with the outside world, so can the cell's contents. The prisoner must be fed, so nutrients must be able to enter every living cell. The cell has a sanitary system for the elimination of waste. And, just as inmates may work to provide materials for society outside the prison, a cell may produce materials for life outside the cell.

Cells come in two types: prokaryotes and eukaryotes. (Viruses also bear some similarities to cells, but these are limited. In fact, many scientists don't consider viruses "living.") Prokaryotic cells are the simplest type of cells. Many one-celled organisms are prokaryotes.

The simplest way to distinguish between these two types of cells is that a *prokaryotic cell* contains no well-defined nucleus, whereas the opposite is true for a *eukaryotic cell.*

Prokaryotes

Prokaryotes are mostly bacteria. Besides the lack of a nucleus, a prokaryotic cell has few well-defined structures. The prison wall has three components: a cell wall, an outer membrane, and a plasma membrane. This wall allows a controlled passage of material into and out of the cell. The materials necessary for proper functioning of the cell float about inside it, in a soup known as the *cytoplasm.* Figure 1-1 depicts a simplified version of a prokaryotic cell.

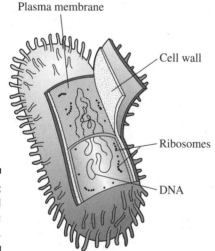

Plasma membrane

Cell wall

Ribosomes

DNA

Figure 1-1:
Simplified
prokaryotic
cell.

Eukaryotes

Eukaryotes are animals, plants, fungi, and *protists* (any organism that isn't a plant, animal, or fungus; many are unicellular organisms, while others are multicellular, like algae). *You* are a eukaryote. In addition to having a nucleus, eukaryotic cells have a number of membrane-enclosed components known as *organelles.* Eukaryotic organisms may be either unicellular or multicellular. In general, eukaryotic cells contain much more genetic material than prokaryotic cells.

Animal Cells and How They Work

All animal cells (which, as you now know, are eukaryotic cells) have a number of components, most of which are considered to be organelles. These components, and a few others, are also present in plant cells (see the section "A Brief Look at Plant Cells" later in the chapter). Figure 1-2 illustrates a simplified animal cell.

The primary components of animal cells include

✔ **Plasma membrane:** This separates the material inside the cell from everything outside the cell. The *plasma* or *cytoplasm* is the fluid inside the cell. For the sake of the cell's health, this fluid shouldn't leak out. However, necessary materials must be able to enter through the membrane, and other materials, including waste, must be able to exit through the membrane. (Imagine what a cesspool that cell would become if the waste products couldn't get out!)

Transport through the membrane may be active or passive. *Active transport* requires that a price be paid for a ticket to enter (or leave) the cell. The cost of the ticket is energy. *Passive transport* doesn't require a ticket. Passive transport methods include *diffusion, osmosis,* and *filtration.*

✔ **Centrioles:** These behave as the cell's "train conductors." They organize structural components of the cell like *microtubules,* which help move the cell's parts during cell division.

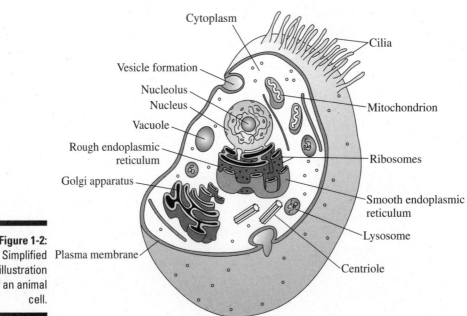

Figure 1-2:
Simplified illustration of an animal cell.

Cytoplasm

Cilia

Vesicle formation

Nucleolus

Nucleus

Mitochondrion

Vacuole

Rough endoplasmic reticulum

Ribosomes

Golgi apparatus

Smooth endoplasmic reticulum

Lysosome

Plasma membrane

Centriole

✔ **Endoplasmic reticulum:** The cell can be thought of as a smoothly running factory. The *endoplasmic reticulum* is the main part of the cell factory. This structure has two basic regions, known as the *rough* endoplasmic reticulum, which contains *ribosomes,* and the *smooth* endoplasmic reticulum, which does not (find out more about ribosomes and their function later in this list). The rough endoplasmic reticulum, through the ribosomes, is the factory's assembly line. The smooth endoplasmic reticulum is more like the shipping department, which ships the products of the reactions that occur within the cell to the Golgi apparatus.

✔ **Golgi apparatus:** This structure serves as the cell's postal system. It looks a bit like a maze, and within it, materials produced by the cell are packaged in *vesicles* — small, membrane-enclosed sacs. The vesicles are then mailed to other organelles or to the cell membrane for export. The cell membrane contains "customs officers" (called *channels*) that allow secretion of the contents from the cell. Secreted substances are then available for other cells or organs.

✔ **Lysosomes:** These are the cell's landfills. They contain digestive enzymes that break down substances that may harm the cell (Chapter 6 has a lot more about enzymes). The products of this digestion may then safely move out of the lysosomes and back into the cell. Lysosomes also digest "dead" organelles. This slightly disturbing process, called *autodigestion,* is really part of the cell digesting itself. (We've never gotten *that* hungry!)

✔ **Mitochondria:** These structures are the cell's power plants, where the cell produces energy. Mitochondria (singular *mitochondrion*) use food, primarily the carbohydrate *glucose,* to produce energy, which comes mainly from breaking down *adenosine triphosphate* (or ATP, to which Chapter 13 is dedicated).

✔ **Nucleus/nucleolus:** Each cell has a *nucleus* and, inside it, a *nucleolus.* These serve as the cell's control center and are the root from which all future generations originate. A double layer known as the *nuclear membrane* surrounds the nucleus. Usually the nucleus contains a mass of material called *chromatin.* If the cell is entering a stage leading to reproducing itself through cell division, the chromatin separates into *chromosomes.*

In addition to conveying genetic information to future generations, the nucleus produces two important molecules for the interpretation of this information. These molecules are *messenger ribonucleic acid* (mRNA) and *transfer ribonucleic acid* (tRNA). The nucleolus produces a third type of ribonucleic acid known as *ribosomal ribonucleic acid* (rRNA). (Chapter 9 is all about nucleic acids.)

✔ **Ribosomes:** These contain protein and ribonucleic acid subunits. In the ribosomes, the amino acids are assembled into *proteins.* Many of these proteins are enzymes, which are part of nearly every process that occurs in the organism. (Part II of this book is devoted to amino acids, proteins, and enzymes.)

✔ **Small vacuoles:** Also known as simply *vacuoles,* these serve a variety of functions, including storage and transport of materials. The stored materials may be for later use or may be waste material that the cell no longer needs.

A Brief Look at Plant Cells

Plant cells contain the same components as animal cells, plus a cell wall, a large vacuole, and, in the case of green plants, chloroplasts. Figure 1-3 illustrates a typical plant cell.

The *cell wall* is composed of *cellulose.* Cellulose, like starch, is a polymer of glucose. The cell wall provides structure and rigidity.

The *large vacuole* serves as a warehouse for large starch molecules. Glucose, which is produced by photosynthesis, is converted to *starch,* a polymer of glucose. At some later time, this starch is available as an energy source. (Chapter 7 talks a lot more about glucose and other carbohydrates.)

Chloroplasts, present in green plants, are specialized chemical factories. These are the sites of *photosynthesis,* in which *chlorophyll* absorbs sunlight and uses this energy to combine carbon dioxide and water to produce glucose and release oxygen gas.

The green color of many plant leaves is due to the magnesium-containing compound chlorophyll.

Now that you know a little about cells, press on and let's do some biochemistry!

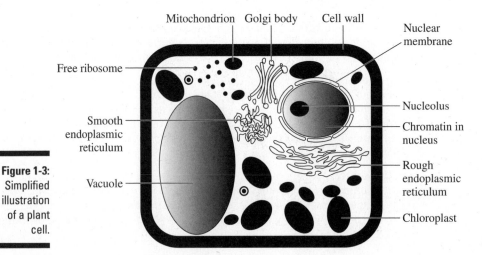

Figure 1-3:
Simplified
illustration
of a plant
cell.

Mitochondrion   Golgi body   Cell wall

Nuclear membrane

Free ribosome

Smooth endoplasmic reticulum

Nucleolus

Chromatin in nucleus

Vacuole

Rough endoplasmic reticulum

Chloroplast

Chapter 2

Seems So Basic: Water Chemistry and pH

. . . .

In This Chapter

▶ Understanding the roles and properties of water

▶ Exploring the differences between acids and bases

▶ Controlling pH with buffers

. . . .

Water is one of the most important substances on earth. People swim, bathe, boat, and fish in it. It carries waste from people's homes and is used in the generation of electrical power. Humans drink it in a variety of forms: pure water, soft drinks, tea, coffee, beer, and so on. Water, in one form or another, moderates the temperature of the earth and of the human body.

In the area of biochemistry, water is also one of the lead actors. The human body is about 70 percent water. Water plays a role in the transport of material to and from cells. And many, many aqueous solutions take part in the biochemical reactions in the body.

In this chapter, we examine the water molecule's structure and properties. We explain how water behaves as a solvent. We also look at the properties of acids and bases and the equilibria that they may undergo. Finally, we discuss the pH scale and buffers, including the infamous Henderson-Hasselbalch equation. So sit back, grab a glass of water (or your favorite water-based beverage), and dive in!

The Fundamentals of H$_2$O

Water is essential to life; in fact, human beings are essentially big sacks of water. Water accounts for 60 to 95 percent of living human cells, and 55 percent of the water in the human body is in intracellular fluids. The remaining 45 percent (extracellular) is divided among the following:

✔ Plasma (8 percent)

✔ Interstitial (between cells) and lymph (22 percent)

✔ Connective tissue, cartilage, and bone (15 percent)

Water also is necessary as a solvent for the multitude of biochemical reactions that occur in the body:

✔ Water acts as a transport medium across membranes, carrying substances into and out of cells.

✔ Water helps maintain body temperature.

✔ Water acts as a solvent (carrying dissolved chemicals) in the digestive and waste excretion systems.

Healthy humans have an intake/loss of about 2 liters of water per day. The intake is about 45 percent from liquids and 40 percent from food, with the remainder coming from the oxidation of food. The loss is about 50 percent from urine and 5 percent from feces, with the remainder leaving through evaporation from the skin and lungs. A water balance must be maintained within the body. If the water loss significantly exceeds the intake, the body experiences dehydration. If the intake significantly exceeds the water loss, water builds up in the body and causes *edema* (fluid retention in tissues).

In the following sections, we touch on the basic properties of this must-have liquid, as well as its most important biochemical function.

Let's get wet! The physical properties of water

The medium in which biological systems operate is water, and the physical properties of water influence the biological systems. Therefore, it's important to review some water properties from general chemistry.

Water is a polar molecule

Water is a *bent* molecule, not linear (see Figure 2-1). The hydrogen atoms have a partially positive charge ($\delta+$); the oxygen atom has a partially negative charge ($\delta-$). This charge distribution is due to the *electronegativity* difference between hydrogen and oxygen atoms (the attraction that an atom has for a bonding pair of electrons). The water molecule in Figure 2-1 is shown in its bent shape with a bond angle of about 105 degrees.

Figure 2-1:
Structure
of a water
molecule.

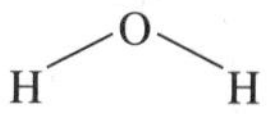

Water has strong intermolecular forces

Normally, partial charges such as those found in a water molecule result in an intermolecular force known as a *dipole-dipole force,* in which the positive end of one molecule attracts the negative end of another molecule. The very high electronegativity of oxygen combined with the fact that a hydrogen atom has only one electron results in a charge difference significantly greater than you'd normally expect. This charge difference leads to stronger-than-expected intermolecular forces, and these forces have a special name: *hydrogen bonds.*

The term *hydrogen bond* doesn't refer to an actual bond to a hydrogen atom but to the overall *interaction* of a hydrogen atom bonded to either oxygen, nitrogen, or fluorine atoms with an oxygen, nitrogen, or fluorine on another molecule *(intermolecular)* or the same molecule *(intramolecular).* Hence the term *intermolecular force.* (Note that although hydrogen bonds occur when hydrogen bonds to fluorine, you don't normally find such combinations in biological systems.)

Hydrogen bonds in oxygen- and nitrogen-containing molecules are very important in biochemistry because they influence reactions between such molecules and the structures of these biological molecules. The interaction between water and other molecules in which there may be an opportunity for hydrogen bonding explains such properties as solubility in water and reactions that occur with water as a solvent (more on that in a minute).

One environmentally important consequence of hydrogen bonding is that, upon freezing, water molecules are held in a solid form that's less dense than the liquid form. The hydrogen bonds lock the water molecules into a crystalline lattice that contains large holes, which decreases the density of the ice. The less-dense ice — whether in the form of an ice cube or an iceberg — floats on liquid water. In nearly all other cases where a solid interacts with water, the reverse is true: The solid sinks in the liquid. So why is the buoyancy of ice important? Ask ice fishermen! The layer of ice that forms on the surface of cold bodies of water insulates the liquid from the cold air, protecting the organisms that live under the ice.

Water has a high specific heat

Specific heat is the amount of heat required to change the temperature of a gram of water 1 degree Celsius. A high specific heat means that changing the temperature of water isn't easy. Water also has a high *heat of vaporization.* Humans can rid their bodies of a great deal of heat when their sweat evaporates from their skin, making sweat a very effective cooling method. We're sure you'll notice this cooling effect during your biochem exams.

Because of water's high specific heat and heat of vaporization, lakes and oceans can absorb and release a large amount of heat without a dramatic change in temperature. This give-and-take helps moderate the earth's temperature and makes it easier for an organism to control its body temperature. Warmblooded animals can maintain a constant temperature, and coldblooded animals — including lawyers and some chemistry teachers — can absorb enough heat during the day to last them through the night.

Water's most important biochemical role: The solvent

The polar nature of water means that it attracts (soaks up) other polar materials. Water is often called the *universal solvent* because it dissolves so many types of substances. Many ionic substances dissolve in water because the negative ends of the water molecules attract the *cations* (positively charged ions) from the *ionic compound* (compound resulting from the reaction of a metal with a nonmetal) and the positive ends attract the *anions* (negatively charged ions). *Covalently bonded* (resulting from the reactions between nonmetals) polar substances, such as alcohols and sugars, also are soluble in water because of the dipole-dipole (or hydrogen-bonding) interactions. However, covalently bonded nonpolar substances, such as fats and oils, aren't soluble in water. Check out *Chemistry For Dummies* (written by this book's coauthor, John T. Moore, and published by Wiley) for a discussion of chemical bonding.

Polar molecules, because of their ability to interact with water molecules, are classified as *hydrophilic* (water-loving). Nonpolar molecules, which don't appreciably interact with (dissolve in) water, are classified as *hydrophobic* (water-hating). Some molecules are *amphipathic* because they have both hydrophilic and hydrophobic regions.

Figure 2-2 shows the structure of a typical amphipathic molecule. The molecule appears on the left, with its hydrophilic and hydrophobic regions shown. The alternate portion of the figure is a symbolic way of representing the molecule. The round "head" is the hydrophilic portion, and the long "tail" is the hydrophobic portion.

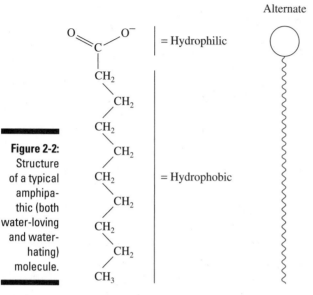

Figure 2-2:
Structure
of a typical
amphipa-
thic (both
water-loving
and water-
hating)
molecule.

Certain amphipathic molecules, such as soap molecules, can form *micelles,* or very tiny droplets that surround insoluble materials. This characteristic is the basis of the cleaning power of soaps and detergents. The hydrophobic portion of the molecule (a long hydrocarbon chain) dissolves in a nonpolar substance, such as normally insoluble grease and oil, leaving the hydrophilic portion (commonly an ionic end) out in the water. Soap or detergent breaks up the grease or oil and keeps it in solution so it can go down the drain.

A micelle behaves as a large polar molecule (see Figure 2-3). The structure of a micelle is closely related to the structure of cell membranes.

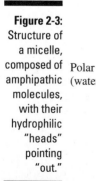

Figure 2-3:
Structure of
a micelle,
composed of
amphipathic
molecules,
with their
hydrophilic
"heads"
pointing
"out."

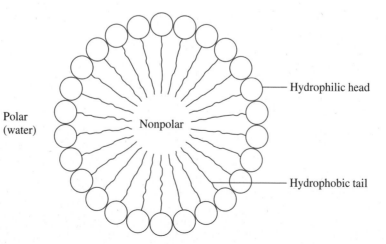

Hydrogen Ion Concentration: Acids and Bases

In aqueous solutions — especially in biological systems — the concentration of hydrogen ions (H^+) is very important. Biological systems often take great pains to make sure that their hydrogen ion concentration — represented as $[H^+]$ or by the measurement of pH (the measure of acidity in a solution) — doesn't change.

Even minor changes in hydrogen ion concentration can have dire consequences to a living organism. For example, in human blood, only a very small range of hydrogen ions allows the body to function properly. Hydrogen ion concentrations higher or lower than this range can cause death.

Because living organisms are so dependent on pH, we review the concepts of acids, bases, and pH in the following sections.

Achieving equilibrium

When the concentrations of hydrogen ion (H^+) and hydroxide ion (OH^-) are the same, a solution is *neutral*. If the hydrogen ion concentration exceeds the hydroxide ion concentration, the solution is *acidic*. If the hydroxide ion concentration is greater, the solution is *basic*. These chemical species are related through a chemical equilibrium.

Acidic solutions, such as lemon juice, taste sour. Basic solutions, such as tonic water, taste bitter. (The addition of gin doesn't change the bitter taste!)

The equilibrium of hydrogen ions is present in all aqueous solutions. Water may or may not be the major hydrogen ion source (usually it isn't). Water is a contributor to the hydrogen ion concentration because it undergoes *autoionization*, as shown by the following equation:

$$H_2O(l) \rightleftarrows H^+(aq) + OH^-(aq)$$

You often see $H^+(aq)$ represented as H_3O^+.

The double arrow ($\rightleftarrows$) indicates that this reaction (represented by the equation) is an equilibrium; as such, there must be an associated *equilibrium constant* (K). The equilibrium constant in the preceding equation is K_w. The value of K_w is the product of the concentrations of the hydrogen ion and the hydroxide ion:

$$K_w = [H^+] [OH^-] = 1.0 \times 10^{-14} \text{ (at } 25°C)$$

The value of the constant K_w, like all K's, is only constant if the temperature is constant. In the human body, where T = 37°C, $K_w = 2.4 \times 10^{-14}$.

In pure water, at 25°C, $[H^+] = 1.0 \times 10^{-7}$ M (1.6×10^{-7} M at 37°C). The hydroxide ion concentration is the same as the hydrogen ion concentration because they're formed in equal amounts during the autoionization reaction. Keep in mind that $[H^+] = [OH^-]$ only in pure water.

M is a concentration term, the molarity. *Molarity* is the number of moles of solute per liter of solution.

Understanding the pH scale

Expressing hydrogen ion concentrations in an exponential form, such as 1.0×10^{-7}, isn't always convenient. Thankfully, you have a way of simplifying the representation of the hydrogen ion concentration: the pH. You can calculate the pH for any solution by using the following equation:

$$pH = -\log [H^+]$$

For instance, in the case of a solution with a hydrogen ion concentration of 1.0×10^{-7} M, the pH would be

$$pH = -\log (1.0 \times 10^{-7}) = 7.0$$

Table 2-1 gives similar calculations for some hydrogen ion concentrations.

Table 2-1	The pH Scale and the Associated Hydrogen Ion Concentration	
[H⁺]	*pH*	*Solution Property*
1.0×10^{0} M	0	Acidic
1.0×10^{-1} M	1	Acidic
1.0×10^{-2} M	2	Acidic
1.0×10^{-3} M	3	Acidic
1.0×10^{-4} M	4	Acidic
1.0×10^{-5} M	5	Acidic
1.0×10^{-6} M	6	Acidic
1.0×10^{-7} M	7	Neutral
1.0×10^{-8} M	8	Basic
1.0×10^{-9} M	9	Basic

(continued)

Table 2-1 *(continued)*

[H⁺]	pH	Solution Property
1.0×10^{-10} M	10	Basic
1.0×10^{-11} M	11	Basic
1.0×10^{-12} M	12	Basic
1.0×10^{-13} M	13	Basic
1.0×10^{-14} M	14	Basic

Solutions with a pH less than 7 are acidic. Solutions with a pH greater than 7 are basic. Solutions whose pH is 7 are neutral. The pH of pure water is 7. Be careful, though: Not every solution that has a pH of 7 is pure water! For example, if you add table salt to water, the pH remains at 7, but the resulting solution is certainly not pure water.

The pH scale is an open-ended scale, meaning that a solution can have a pH greater than 14 or less than 0. For example, the pH of a 1.0×10^{1} M solution of hydrochloric acid is –1. John loves to ask questions based on this topic to his advanced chemistry students! The 0–14 scale is a convenient part of the pH scale for most real-world solutions — especially ones found in biochemistry. Most biological systems have a pH near 7, although significant deviations may exist (for example, the pH in your stomach is close to 1).

Calculating pOH

You can calculate *pOH* (related to the concentration of the hydroxide ion) in a similar manner to the pH calculation. That is, you can use the equation pOH = –log [OH⁻]. You can calculate the hydroxide ion concentration from the hydrogen ion concentration and the K_w (equilibrium constant) relationship:

$$[OH^-] = K_w \div [H^+]$$

A useful shortcut to get from pH to pOH is the following relationship:
pH + pOH = 14.00 for any aqueous solution ($14.00 = pK_w = -\log K_w = -\log 1.0 \times 10^{-14}$).

For example, if a solution has a $[H^+] = 6.2 \times 10^{-6}$, its pH would be

$$pH = -\log [H^+]$$
$$pH = -\log [6.2 \times 10^{-6}]$$
$$pH = 5.21$$

The calculation for the pOH of that solution is pretty simple:

pOH = 14.00 – pH = 14.00 – 5.21 = 8.79

Now, if you have the pH or pOH, getting the corresponding [H⁺] or [OH⁻] is a simple task:

$[H^+] = 10^{-pH}$ and $[OH^-] = 10^{-pOH}$

For example, a solution with a pH of 7.35 has a $[H^+] = 10^{-7.35} = 2.2 \times 10^{-7}$.

Applying the Brønsted-Lowry theory

Because the acidity (pH) of the biological medium is so very important, here we take a look at one of the most accepted theories concerning acids and bases — the Brønsted-Lowry theory. According to this theory, acids are proton (H⁺) donors, and bases are proton acceptors.

Swapping hydrogens between acids and bases

Acids increase the hydrogen ion concentration of a solution (they lower the pH, in other words). Some acids, known as *strong acids,* are very efficient at changing hydrogen ion concentration; they essentially completely ionize in water. Most acids — particularly biologically important acids — aren't very efficient at generating hydrogen ions; they only partially ionize in water. These acids are known as *weak acids.*

Bases accept (react with) rather than donate hydrogen ions in solutions. Bases decrease the hydrogen ion concentration in solutions because they react with these ions. Strong bases, although they can accept hydrogen ions very well, aren't too important in biological systems. The majority of biologically important bases are weak bases.

 The Brønsted-Lowry theory helps to explain the behavior of acids and bases with respect to equilibrium. A Brønsted-Lowry acid is a hydrogen ion (H⁺) donor, and a Brønsted-Lowry base is a hydrogen ion acceptor. Acetic acid, a weak acid found in vinegar, partially ionizes in solution, evidenced by the following equation:

$CH_3COOH \rightleftharpoons H^+ + CH_3COO^-$

The double arrow indicates that the acetic acid doesn't completely ionize. (For a strong acid, complete ionization would occur, indicated by a single arrow.) The equilibrium arrow ($\rightleftharpoons$) indicates that all three chemical species are present in the solution: the acetic acid, the acetate ion, and the hydrogen ion, along with the water solvent.

In the Brønsted-Lowry theory, you consider the acetate ion to be a base because it can accept a hydrogen ion to become acetic acid. According to this theory, two substances differing by only one hydrogen ion (H^+) — such as acetic acid and the acetate ion — are members of a *conjugate acid-base pair*. The species with one additional hydrogen ion is the *conjugate acid* (CA), and the species with one less hydrogen ion is the *conjugate base* (CB).

You can express the equilibrium from the acetate example, like all equilibria, by using a *mass-action expression* — as long as a balance among the species is present. This expression is also known as a *reaction quotient* or an *equilibrium constant.* For acetic acid, this expression is as follows:

$$K_a = \frac{\left[H^+\right]\left[CH_3COO^-\right]}{\left[CH_3COOH\right]}$$

The *a* subscript means that this expression represents an acid. The square brackets refer to the molar equilibrium concentrations of the species present. You can express the K_a as a pK_a. The calculation of pK_a is similar to the calculation of pH:

$$pK_a = -\log K_a$$

In terms of conjugate acids and bases, every K_a expression appears as

$$K_a = \frac{\left[H^+\right]\left[CB\right]}{\left[CA\right]}$$

No variations are allowed in this equation other than the actual formulas of the conjugate acid and base.

Like an acid, a base has a K_b value (the subscript *b* meaning *base*). A weak base, like ammonia, is part of the following equilibrium:

$$NH_3 + H_2O \rightleftharpoons OH^- + NH_4^+$$

The equilibrium constant expression for this equilibrium is

$$K_b = \frac{\left[OH^-\right]\left[NH_4^+\right]}{\left[NH_3\right]}$$

The generic form of a K_b expression is

$$K_b = \frac{\left[OH^-\right]\left[CA\right]}{\left[CB\right]}$$

As with a K_a expression, a K_b expression has no variations other than the actual formulas of the conjugate acid and base.

Every acid has a K_a, and its corresponding conjugate base has a K_b. The K_a and the K_b of a conjugate acid-base pair are related by the K_w — the ionization constant for water. For a conjugate acid-base pair, $K_a K_b = K_w = 1.0 \times 10^{-14}$. In addition, you can use the following shortcut: $pK_a + pK_b = 14.00$. (Are you enjoying this little bit of math? Biochemistry has a little more math than organic chemistry, but not nearly as much as general chemistry, so hang in there!)

The K_b for the acetate ion, the conjugate base of acetic acid, would be associated with the following equilibrium expression:

$$CH_3COO^- + H_2O \rightleftarrows OH^- + CH_3COOH$$

The K_a for the ammonium ion, the conjugate acid of ammonia, would be associated with the following equilibrium expression:

$$NH_4^+ \rightleftarrows H^+ + NH_3$$

An acid may be capable of donating more than one hydrogen ion. A biologically important example of this type of acid is phosphoric acid (H_3PO_4), which is a *triprotic* acid (meaning it can donate three hydrogen ions, one at a time). The equilibria for this acid are

$$K_{a1}: H_3PO_4 \rightleftarrows H^+ + H_2PO_4^-$$
$$K_{a2}: H_2PO_4^- \rightleftarrows H^+ + HPO_4^{2-}$$
$$K_{a3}: HPO_4^{2-} \rightleftarrows H^+ + PO_4^{3-}$$

The subscripts are modified to indicate the loss of hydrogen 1, hydrogen 2, or hydrogen 3. The associated K_a expressions are all of the form

$$K_a = \frac{[H^+][CB]}{[CA]}$$

Here's the breakdown for each K_a:

$$K_{a_1} = \frac{[H^+][H_2PO_4^-]}{[H_3PO_4]}$$

$$K_{a_2} = \frac{[H^+][HPO_4^{2-}]}{[H_2PO_4^-]}$$

$$K_{a_3} = \frac{[H^+][PO_4^{3-}]}{[HPO_4^{2-}]}$$

The value for each successive equilibrium constant often is significantly lower than the preceding value. Table 2-2 illustrates some biologically important acids. You can refer to this table when working buffer problems or determining which acid is stronger.

Table 2-2	The K_a Values for Biologically Important Acids		
Acid	**K_{a1}**	**K_{a2}**	**K_{a3}**
Acetic acid (CH_3COOH)	1.7×10^{-5}		
Pyruvic acid ($CH_3COCOOH$)	3.2×10^{-3}		
Lactic acid ($CH_3CHOHCOOH$)	1.4×10^{-4}		
Succinic acid ($HOOCCH_2CH_2COOH$)	6.2×10^{-5}	2.3×10^{-6}	
Carbonic acid (H_2CO_3)	4.5×10^{-7}	5.0×10^{-11}	
Citric acid ($HOOCCH_2C(OH)(COOH)-CH_2COOH$)	8.1×10^{-4}	1.8×10^{-5}	3.9×10^{-6}
Phosphoric acid (H_3PO_4)	7.6×10^{-3}	6.2×10^{-8}	2.2×10^{-13}

Acting as either an acid or a base

Some substances can't make up their minds about what they are; they can act as either an acid or a base. Chemists classify these substances as *amphiprotic* or *amphoteric* substances. For example, the bicarbonate ion (HCO_3^-) can act as either an acid or a base:

$$HCO_3^- \rightleftarrows H^+ + CO_3^{2-}$$

$$HCO_3^- + H_2O \rightleftarrows OH^- + H_2CO_3$$

Biochemically important molecules may also exhibit amphiprotic behavior. Amino acids contain both a basic amine ($-NH_2$) group and an acidic carboxyl ($-COOH$) group. Therefore, they can act as either acids or bases. For example, glycine (H_2N-CH_2-COOH) may undergo the following reactions:

$$H_2N-CH_2-COOH \rightleftarrows H^+ + H_2N-CH_2-COO^-$$

$$H_2N-CH_2-COOH + H_2O \rightleftarrows OH^- + {}^+H_3N-CH_2-COOH$$

In fact, amino acids may undergo proton transfer from the carboxyl end to the amine end, forming an overall neutral species that has a positive and negative end. Species such as these are called *zwitterions* (not to be confused with twitterions — people who tweet until their thumbs fall off):

$$H_2N-CH_2-COOH \rightleftarrows {}^+H_3N-CH_2-COO^-$$

Buffers and pH Control

A solution that contains the conjugate acid-base pair of any weak acid or base in relative proportions to resist pH change when small amounts of either an acid or a base are added is a *buffer solution*. Therefore, buffers control the pH of the solution. Buffer solutions are important in most biological systems. Many biological processes proceed effectively only within a limited pH range. The presence of buffer systems keeps the pH within this limited range.

Identifying common physiological buffers

In the human body, the pH of various body fluids is important. The pH of blood is 7.4, the pH of stomach acid is 1–2, and the pH in the intestinal tract is 8–9. If the pH of blood is more than 0.2 pH units lower than normal, a condition known as *acidosis* results; a corresponding increase in pH of about the same magnitude is *alkalosis*. Acidosis and alkalosis, which may lead to serious health problems, each have two general causes:

- *Respiratory acidosis* is the result of many diseases that impair respiration, including pneumonia, emphysema, and asthma. These diseases are marked by inefficient expulsion of carbon dioxide, leading to an increase in the concentration of carbonic acid, H_2CO_3.

- *Metabolic acidosis* is due to a decrease in the concentration of HCO_3^- (the bicarbonate ion). This decrease may be the result of certain kidney diseases, uncontrolled diabetes, and cases of vomiting involving nonacid fluids. Poisoning by an acid salt may also lead to metabolic acidosis.

- *Respiratory alkalosis* may result from hyperventilation, because this excessive removal of carbon dioxide can lead to a decrease in the H_2CO_3 concentration. Immediate treatment includes breathing into a paper bag, which increases the carbon dioxide concentration in the inhaled air and, therefore, in the blood.

- *Metabolic alkalosis* may result from excessive vomiting of stomach acid.

To resist these pH problems, the blood has a number of buffer systems — systems that resist a change in pH by reacting with either added acids or bases. In general, buffers may be amphiprotic substances or mixtures of weak acids and weak bases. In the body these include several proteins in blood plasma and the bicarbonate buffer system.

The *bicarbonate buffer system* is the main extracellular buffer system. This system also provides a means of eliminating carbon dioxide. The dissolution of carbon dioxide in aqueous systems sets up the following equilibrium:

$$CO_2 + H_2O \rightleftarrows H_2CO_3 \rightleftarrows H^+ + HCO_3^-$$

The presence of the conjugate acid-base pair (H_2CO_3 and HCO_3^-) means that this is a buffer system. The conjugate acid-base ratio is about 20:1 at a pH of 7.4 in the bloodstream. This buffer system is coupled with the following equilibrium (instrumental in the removal of carbon dioxide in the lungs):

$$CO_2(\text{blood}) \rightleftarrows CO_2(\text{lungs})$$

The second ionization of phosphoric acid, K_{a2}, is the primary intracellular buffer system. The pH of this conjugate acid-base pair ($H_2PO_4^-$ and HPO_4^{2-}) is 7.21 for a solution with equal concentrations of these two species.

Calculating a buffer's pH

To determine a buffer's pH, you may use a K_a or K_b calculation, as we discuss earlier in the chapter, or you may use the Henderson-Hasselbalch equation, which is a shortcut.

The Henderson-Hasselbalch equation takes two forms:

$$pH = pK_a + \log\frac{[CB]}{[CA]}$$

and

$$pOH = pK_b + \log\frac{[CA]}{[CB]}$$

The terms in either form are the same as those we define earlier in the chapter. For example, suppose you want to calculate the pH of a buffer composed of 0.15 M pyruvic acid and 0.25 M sodium pyruvate. Referring back to Table 2-2, you see that the K_a of pyruvic acid is 3.2×10^{-3}.

The pK_a would be 2.50. Therefore:

$$pH = pK_a + \log\frac{[CB]}{[CA]}$$

$$pH = -\log 3.2 \times 10^{-3} + \log\frac{[CH_3COCOO^-]}{[CH_3COCOOH]}$$

$$pH = 2.50 + \log\frac{[0.25]}{[0.15]}$$

$$pH = 2.50 + \log(1.67)$$

$$pH = 2.50 + 0.22 = 2.72$$

The greater the values of [CA] and [CB], the greater the buffer capacity of the solution. The buffer capacity indicates how much acid or base may be added to a buffer before the buffer ceases to function. A buffer in which the [CA] = [CB] = 1.0 would have a much higher buffer capacity for adding either acids or bases than a buffer in which the [CA] = [CB] = 0.1. If there were a buffer in which [CA] = 1.0 and its [CB] = 0.1, the buffer would have a higher buffer capacity for additions of a base than for additions of an acid because the buffer contains more acid than base. For the buffer to be as flexible as possible, the concentrations of the conjugate acid-base pair should be as close to equal as possible and as high as possible.

Chapter 3

Fun with Carbon: Organic Chemistry

*M*ost biologically important molecules are composed of *organic compounds,* compounds of carbon. Therefore, you, as a student of biochemistry, must have a general knowledge of organic chemistry, which is the study of carbon compounds, in order to understand the functions and reactions of biochemical molecules. In this chapter we review the basics of organic chemistry, including the various functional groups and isomers that are important in the field of biochemistry. (We're sure that this will bring back fond memories of your organic classes and labs.)

The Role of Carbon in the Study of Life

Long ago, scientists believed that all carbon compounds were the result of biological processes, which meant that organic chemistry was synonymous with biochemistry under what was known as the *Vital Force theory.* In the mid-1800s, though, researchers such as Friedrich Wöhler debunked that long-held notion; the synthesis of urea from an inorganic material (ammonium cyanate, NH_4OCN) showed that other paths to the production of carbon compounds existed. Organic chemists now synthesize many important organic chemicals without the use of living organisms; however, biosynthesis is still an important source of many organic compounds.

Why are there so many carbon compounds? The answer lies primarily in two reasons, both tied to carbon's versatility in creating stable bonds:

✔ **Carbon bonds to itself.** Carbon atoms are capable of forming stable bonds to other carbon atoms. The process of one type of atom bonding to identical atoms is *catenation*. Many other elements can catenate, but carbon is the most effective at it. There appears to be no limit to how many carbon atoms can link together. These linkages may be in chains, branched chains, or rings, as shown in Figure 3-1.

✔ **Carbon bonds to other elements.** Carbon is capable of forming stable bonds to a number of other elements. These include the biochemically important elements hydrogen, nitrogen, oxygen, and sulfur. The latter three elements form the foundation of most of the *functional groups* (reactive groups of a molecule) necessary for life. Bonds between carbon and hydrogen are usually unreactive under biochemical conditions; thus, hydrogen often serves as an "inert" substituent.

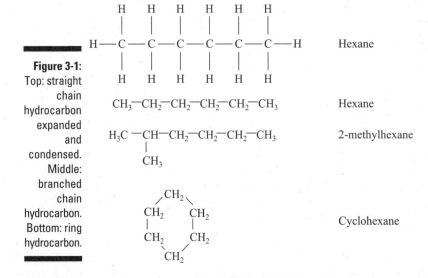

Figure 3-1:
Top: straight chain hydrocarbon expanded and condensed. Middle: branched chain hydrocarbon. Bottom: ring hydrocarbon.

Hexane

Hexane

2-methylhexane

Cyclohexane

It's All in the Numbers: Carbon Bonds

Carbon is capable of forming four bonds. In bonding to itself and other elements, carbon uses a variety of types of *hybridization*. When it bonds to another carbon molecule, for example, possible hybridizations include four single bonds, one double and two single bonds, two double bonds, or a triple and a single bond. Double bonds to oxygen atoms are particularly important in many biochemicals. Table 3-1 shows the number of bonds carbon may form with some selected nonmetals, along with the hybridization of those bonds.

Table 3-1	Possible Bonds of Carbon and Selected Nonmetals	
Element	*Number of Possible Bonds with Carbon*	*Some Possible Hybridizations for Second Period Elements*
Carbon (C)	4	4 single (sp^3); 2 single and 1 double (sp^2); 1 single and 1 triple (sp); 2 double (sp)
Nitrogen (N)	3	3 single (sp^3); 1 single and 1 double (sp^2); 1 triple (sp)
Oxygen (O)	2	2 single (sp^3); 1 double (sp^2)
Sulfur (S)	2	2 single (sp^3); 1 double (sp^2)
Hydrogen (H)	1	1 single
Fluorine (F)	1	1 single
Chlorine (Cl)	1	1 single
Bromine (Br)	1	1 single
Iodine (I)	1	1 single

When Forces Attract: Bond Strengths

Covalent bonds are important *intramolecular forces* (forces within the same molecule) in biochemistry. *Intermolecular forces* (forces between chemical species) are also extremely important. Among other things, intermolecular forces are important to *hydrophilic* (water-loving) and *hydrophobic* (water-hating) interactions. (*Phobias* involve fear or hate, so *hydrophobic* is water hating.)

Everybody has 'em: Intermolecular forces

All intermolecular forces are *van der Waals forces* — that is, they're not true bonds in the sense of sharing or transferring electrons but are weaker attractive forces. These forces include London dispersion forces, dipole-dipole forces, hydrogen bonding, and ionic interactions.

London dispersion forces

London dispersion forces are very weak and short-lived attractions between molecules that arise from the nucleus of one atom attracting the electron cloud of another atom. These forces are only significant when other intermolecular forces are not present, as in nonpolar molecules.

Dipole-dipole forces

Dipole-dipole forces exist between polar regions of different molecules. The presence of a dipole means that the molecule has a partially positive ($\delta+$) end and a partially negative ($\delta-$) end. Oppositely charged partial charges attract each other, whereas like partial charges repel. In most cases, biological systems utilize a special type of dipole-dipole force known as *hydrogen bonding* (see the next section). (We said "hydrogen bond," not "hydrogen bomb" — there's a big difference!)

Hydrogen bonding

Hydrogen bonding, as the name implies, involves hydrogen. The hydrogen atom must be bonded to either an oxygen atom or a nitrogen atom. (In non-biological situations, hydrogen bonding also occurs when a hydrogen atom bonds to a fluorine atom.) Hydrogen bonding is significantly stronger than a "normal" dipole-dipole force and is much stronger than London dispersion forces. The hydrogen that bonds to either a nitrogen or an oxygen atom is strongly attracted to a different nitrogen or oxygen atom. Hydrogen bonding may be either intramolecular or intermolecular.

Ionic interactions

In biological systems, *ionic interactions* may serve as intermolecular or intramolecular forces. In some cases, these may involve *metal cations,* such as Na^+, or *anions,* such as Cl^-. In many cases, the cation is an ammonium ion from an amino group, such as RNH_3^+; the anion may be from a carboxylic acid, such as $RCOO^-$. Oppositely charged ions attract each other strongly.

Water-related interactions: Both the lovers and the haters

The predominant factor leading to hydrophobic interactions is the presence of portions of a molecule containing only carbon and hydrogen. Hydrocarbon regions are nonpolar and are attracted to other nonpolar regions by London dispersion forces.

In general, the presence of any atom other than carbon and hydrogen makes a region polar. Oxygen and nitrogen are the most effective elements in biochemistry for making a region of a molecule polar. Sulfur is the least effective of the common biologically important elements at imparting polar character. Dipole-dipole, hydrogen bonding, and ionic interactions are all hydrophilic interactions. London dispersion forces are hydrophobic interactions.

The more carbon and hydrogen atoms that are present, without other atoms, the more important the hydrophobic nature of a region becomes in defining the molecule's properties. Note that a molecule may have both a hydrophilic and a hydrophobic region, and both regions are important to the molecule's behavior. The formation of a *micelle* (see Chapter 2) is an example of using molecules with both hydrophilic and hydrophobic regions. Think of these micelles every time you wash dishes. The soap or detergent dissolves the grease or oil and forms a micelle, keeping the grease in solution so that it can go down the drain.

How bond strengths affect physical properties of substances

The physical properties of biological substances depend on the intermolecular forces present. The order of strength is:

ions > hydrogen bonding > dipole-dipole > London

The strongest types of intermolecular forces involve ions. Next strongest is hydrogen bonding. Polar substances interact through dipole-dipole forces, which are weaker than hydrogen bonds. All biological substances containing oxygen, nitrogen, sulfur, or phosphorus are polar. London forces, the weakest intermolecular forces, are important in nonpolar situations. The hydrocarbon portion of biological molecules is nonpolar.

Melting points, boiling points, and solubility

As the strength of forces decreases, so do the melting points, boiling points, and solubility in water. In addition, the vapor pressure and the solubility in nonpolar solvents increase.

Substances that have a high solubility in water are hydrophilic, and substances that have a low solubility in water are hydrophobic.

A molecule may have both hydrophilic and hydrophobic regions. The region that represents a greater portion of the molecule predominates. For this reason, for example,

CH_3COOH

is more hydrophilic than

$CH_3CH_2CH_2CH_2CH_2 CH_2CH_2CH_2CH_2CH_2COOH$

because the hydrophilic end (-COOH) is a much more significant portion of the entire molecule in the first case than in the second case.

In addition,

$$HOCH_2CH_2CH_2CH_2CH_2CH_2OH$$

is more hydrophilic than

$$CH_3CH_2CH_2CH_2CH_2CH_2OH$$

because of the presence of the second hydrophilic region (-OH).

Odors

Many functional groups have distinctive odors. Small carboxylic acids smell like acetic acid (vinegar), while larger ones have unpleasant odors. Most esters, if volatile, have pleasant odors, which is why esters are used extensively in the food and perfume industry. Most sulfur-containing compounds have strong, unpleasant odors. Small amines have an ammonia odor, whereas larger amines have a fishy odor or worse. That's why people squeeze lemon juice on fish — the acidic lemon juice reacts with the basic amines to form an ammonium salt that doesn't have an odor. (Believe us, there are odors worse than fish — cadaverine is one!)

Getting a Reaction out of a Molecule: Functional Groups

Most carbon compounds have one or more reactive sites composed of a specific grouping of atoms in their structure. These sites are where chemical reactions occur. These specific groupings of atoms that react are called *functional groups*. Functional groups contain atoms other than carbon and hydrogen and/or double or triple bonds, and they define the reactivity of the organic molecule.

Hydrocarbons

Alkanes are *hydrocarbons* — compounds containing only carbon and hydrogen, with no traditional functional groups. For this reason, they aren't very reactive. *Alkenes* and *alkynes* are also hydrocarbons. They contain carbon-carbon double and triple bonds, respectively. The presence of more than one type of bond makes them more reactive. *Aromatic hydrocarbons,* normally ring structures with alternating single and double carbon-to-carbon bonds, contain one or more *aromatic systems,* which are much less reactive than other systems containing double bonds. Alkynes aren't very common in biological systems. Figure 3-2 shows the structure of these compounds.

Figure 3-2:
Examples
of alkanes,
alkenes,
alkynes, and
aromatic
hydrocarbons.

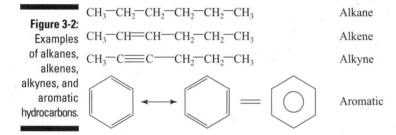

$CH_3-CH_2-CH_2-CH_2-CH_2-CH_3$ Alkane

$CH_3-CH=CH-CH_2-CH_2-CH_3$ Alkene

$CH_3-C\equiv C-CH_2-CH_2-CH_3$ Alkyne

Aromatic

Functional groups with oxygen and sulfur

Many functional groups contain oxygen, including *alcohols, ethers, aldehydes,* and *ketones.* You encounter many of these oxygen-containing functional groups when you study carbohydrates (one of our favorite things). In carbohydrates, many times the ether groups are referred to as *glycoside linkages* (more on that in Chapter 7). In addition, carboxylic acids and esters are important functional groups that appear as fatty acids and in fats and oils.

Alcohols and ethers contain only singly bonded oxygen atoms. An alcohol group attached to an aromatic ring is a *phenol.* Aldehyde and ketone functional groups contain only doubly bonded oxygen atoms. Carboxylic acids and esters contain both singly and doubly bonded oxygen atoms. The combination of a carbon atom connected to an oxygen atom by a double bond is a *carbonyl group.*

Sulfur, the element immediately below oxygen on the periodic table, may replace oxygen in both alcohols and ethers to give *thiols* (mercaptans) and *thioethers.* Many of these sulfur-containing compounds really stink! Sulfur may also form a *disulfide,* which has a bond between two sulfur atoms. Figure 3-3 illustrates these compounds.

R = any organic (hydrocarbon) group

R' = any organic group, which may or may not = R

R—OH
Alcohol

R—O—R'
Ether

$$
\begin{array}{c} O \\ \parallel \\ R-C-H \end{array}
$$
Aldehyde

$$
\begin{array}{c} O \\ \parallel \\ R-C-R' \end{array}
$$
Ketone

$$
\begin{array}{c} O \\ \parallel \\ R-C-OH \end{array}
$$
Carboxylic acid

$$
\begin{array}{c} O \\ \parallel \\ R-C-O-R' \end{array}
$$
Ester

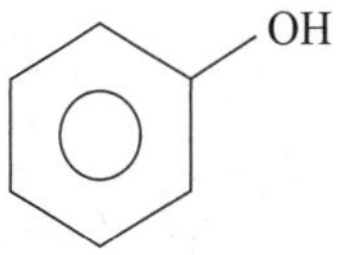

Phenol

R—SH
Thiol

R—S—R'
Thioether

R—S—S—R'
Disulfide

Figure 3-3: Oxygen- and sulfur-containing functional groups.

Functional groups containing nitrogen

Amines and amides are two important functional groups containing nitrogen. *Amines* are present in amino acids and alkaloids. *Amides* are present in proteins, in which they're known as *peptide bonds*.

The difference between amines and amides is that amides have a carbonyl group adjacent to the nitrogen atom. Amines are derivatives of ammonia, NH_3, where one or more organic groups replace hydrogen atoms. In a *primary amine,* an organic group replaces one hydrogen atom. In *secondary* and *tertiary amines,* two or three organic groups, respectively, replace two or three hydrogen atoms. Figure 3-4 shows these compounds, as well as aniline and ammonia.

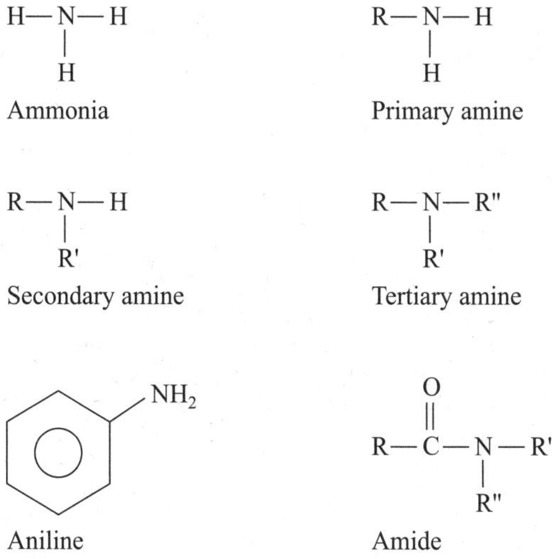

H—N—H
|
H
Ammonia

R—N—H
|
H
Primary amine

R—N—H
|
R'
Secondary amine

R—N—R''
|
R'
Tertiary amine

Figure 3-4:
Some nitrogen-containing functional groups.

NH₂

Aniline

$$O$$
$$\|$$
R—C—N—R'
|
R''
Amide

Alkaloids are basic compounds produced by plants. Examples include nicotine, caffeine, and morphine.

Functional groups containing phosphorus

Phosphorus is also an important element in biological systems and is normally present as part of a *phosphate group.* Phosphate groups come from phosphoric acid, H_3PO_4. The phosphate groups may be alone, part of a diphosphate, part of a triphosphate, or part of a phosphate ester.

Phosphates are in teeth and bone and are a part of the energy transport molecules ATP and ADP (see Chapter 12 for more on these). Figure 3-5 illustrates phosphorous-containing functional groups.

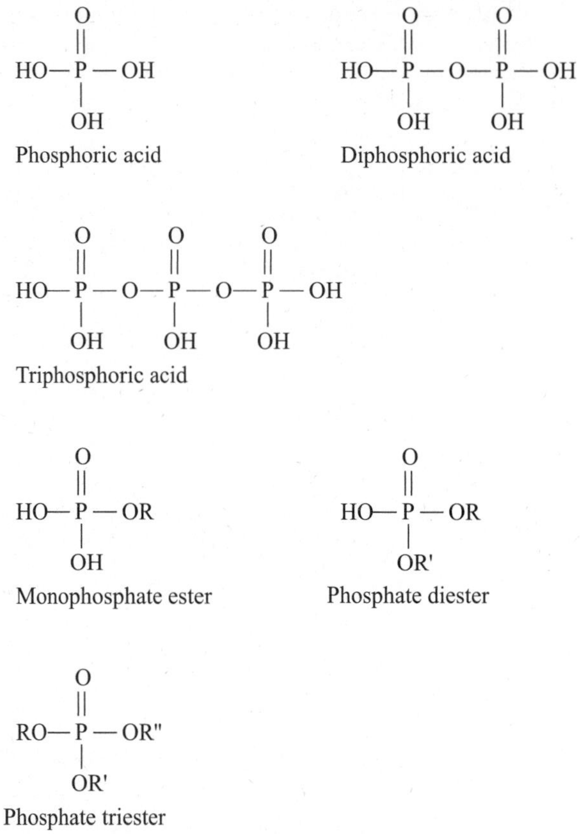

$$
\begin{array}{c}
O \\
\parallel \\
HO-P-OH \\
\mid \\
OH
\end{array}
$$

Phosphoric acid

$$
\begin{array}{c}
O \qquad\quad O \\
\parallel \qquad\quad \parallel \\
HO-P-O-P-OH \\
\mid \qquad\quad \mid \\
OH \qquad\quad OH
\end{array}
$$

Diphosphoric acid

$$
\begin{array}{c}
O \qquad\quad O \qquad\quad O \\
\parallel \qquad\quad \parallel \qquad\quad \parallel \\
HO-P-O-P-O-P-OH \\
\mid \qquad\quad \mid \qquad\quad \mid \\
OH \qquad\quad OH \qquad\quad OH
\end{array}
$$

Triphosphoric acid

$$
\begin{array}{c}
O \\
\parallel \\
HO-P-OR \\
\mid \\
OH
\end{array}
$$

Monophosphate ester

$$
\begin{array}{c}
O \\
\parallel \\
HO-P-OR \\
\mid \\
OR'
\end{array}
$$

Phosphate diester

Figure 3-5:
Phosphorous-
containing
functional
groups.

$$
\begin{array}{c}
O \\
\parallel \\
RO-P-OR'' \\
\mid \\
OR'
\end{array}
$$

Phosphate triester

Reactions of functional groups

As you study the different biochemical molecules and their functions within the living organism, you see that the way a certain molecule reacts is primarily determined by the functional groups in the molecule's structure. Take a few minutes and refresh your organic chemistry knowledge of the typical reactions of the various functional groups.

Alcohols

Alcohols are subject to *oxidation* (loss of electrons, gain of oxygen, loss of hydrogen). Mild oxidation of a primary alcohol (where the –OH is attached to an end carbon) produces an aldehyde, which may undergo further oxidation to a carboxylic acid. Under similar conditions, a secondary alcohol (–OH is attached to a carbon bonded to two other carbons) yields a ketone, and a tertiary alcohol (–OH is attached to a carbon bonded to three other carbons) doesn't react. This behavior is important in the chemistry of many carbohydrates.

The presence of the –OH leads people mistakenly to assume that alcohols are bases. Nothing could be further from the truth! Alcohols, under biological conditions, are neutral compounds. Phenols, in which the –OH is attached to an aromatic ring, though, are weak acids.

Aldehydes and ketones

Aldehydes easily undergo oxidation to carboxylic acids, but ketones don't undergo mild oxidation. With difficulty (unless you use enzymes, biological catalysts), it's possible to reduce aldehydes and ketones back to the appropriate alcohols.

Reducing sugars behave as such because of mild oxidation of the carbonyl groups present. *Tollen's test* uses silver nitrate, which reacts with a reducing sugar to generate a silver mirror on the inside walls of the container. Both *Benedict's test* and *Fehling's test* use copper compounds, and a reducing sugar produces a red precipitate with either test. These simple organic qualitative tests find some use in the biochemical tests we describe later in this book.

The carbonyl group of an aldehyde or ketone may interact with an alcohol to form acetals and hemiacetals. Modern terminology only uses the terms *acetals* and *hemiacetals,* but you may sometimes see the terms *hemiketal,* which is a type of hemiacetal, and *ketal,* a type of acetal. See Figure 3-6 for an illustration of these.

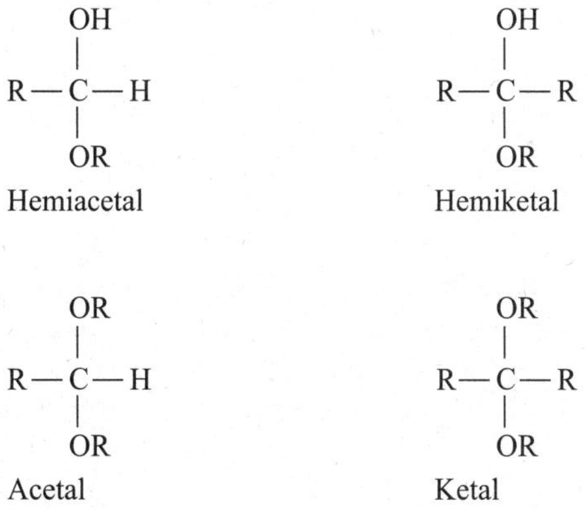

OH | R—C—H | OR Hemiacetal	OH | R—C—R | OR Hemiketal
OR | R—C—H | OR Acetal	OR | R—C—R | OR Ketal

Figure 3-6: Acetals, hemiacetals, hemiketals, and ketals.

Carboxylic acids

Carboxylic acids, along with phosphoric acid, are the most important biological acids. Carboxylic acids react with bases such as the amines to produce salts. The salts contain an ammonium ion from the amine and a carboxylate ion from the acid.

Carboxylic acids combine with alcohols to form esters and can indirectly combine with amines to form amides. Hydrolysis of an ester or an amide breaks the bond and inserts water. An acid, base, or enzyme is needed to catalyze hydrolysis. Under acidic conditions, you can isolate the acid and either the alcohol or the ammonium ion from the amine. Under basic conditions, you can isolate the carboxylate ion and either the alcohol or the amine.

Thiols and amines

Under mild oxidation, two *thiols* join to form a disulfide. Mild reducing conditions, catalyzed by enzymes or through the use of certain reducing agents, reverse this process. Such formation of disulfide linkages is important in the chemistry of many proteins, such as insulin.

Amines are the most important biological bases. As bases, they can react with acids. The behavior is related to the behavior of ammonia.

$$NH_3 + H^+ \text{ (from an acid)} \rightarrow NH_4^+ \text{ (ammonium ion)}$$

$$NRH_2 + H^+ \text{ (from an acid)} \rightarrow NRH_3^+ \text{ (ammonium ion)}$$

$$NR_2H + H^+ \text{ (from an acid)} \rightarrow NR_2H_2^+ \text{ (ammonium ion)}$$

$$NR_3 + H^+ \text{ (from an acid)} \rightarrow NR_3H^+ \text{ (ammonium ion)}$$

Many medications have amine groups. Converting many of these amines to ammonium ions makes them more readily soluble. For example, the reaction of a medication with hydrochloric acid forms the chloride, which often appears on the label as the hydrochloride.

It's possible to replace all the hydrogen atoms from an ammonium ion, NH_4^+, to produce a quaternary ammonium ion, NR_4^+.

Phosphoric acid

Phosphoric acid, H_3PO_4, may behave like a carboxylic acid and form esters. The esters have an organic group, R, replacing one, two, or three of the hydrogen atoms. The resultant compounds are *monoesters, diesters,* and *triesters.* The hydrogen atoms remaining in the monoesters and diesters are acidic.

pH and functional groups

Many of the biological functions of substances are pH-dependent. For this reason, knowing which functional groups are acidic, basic, or neutral is important. Neutral functional groups behave the same no matter what the pH is. Table 3-2 lists the functional groups; whether they're acidic, neutral, or basic; and their weakness level (medium, weak, or very weak). The weaker a substance is in terms of pH, the less likely it's affected by its solution pH.

Table 3-2	Acid-Base Properties of Biologically Important Functional Groups		
Functional Group	*Acid, Base, or Neutral*		*Weakness Level*
Monophosphate esters and diphosphate esters	Acid		Medium
Carboxylic acids	Acid		Weak
Phenols	Acid		Very weak
Thiols	Acid		Very weak
Amine salts	Acid		Very weak
Amines	Base		Weak
Carboxylate ions	Base		Very weak
Alcohols	Neutral		
Carboxylate esters	Neutral		
Ethers	Neutral		
Triphosphate esters	Neutral		
Thioethers	Neutral		
Disulfides	Neutral		
Amides	Neutral		
Ketones	Neutral		
Aldehydes	Neutral		

Same Content, Different Structure: Isomerism

Isomers are compounds that have the same molecular formula but different structural formulas. (It's all in how things are put together.) Some organic and biochemical compounds may exist in different isomeric forms. Many times, especially in biological systems, these different isomers have different properties. The two most common types of isomers in biological systems are cis-trans isomers and isomerism due to the presence of a chiral carbon.

Cis-trans isomers

The presence of carbon-carbon double bonds leads to the possibility of having isomers present. Double bonds are rather restrictive and limit molecular movement. Groups on the same side of the double bond tend to remain in that position *(cis)*, whereas groups on opposite sides tend to remain across the bond from each other *(trans)*. See Figure 3-7 for an illustration of cis and trans isomers.

Figure 3-7:
Cis and trans isomers.

$$\underset{\text{Cl}}{\overset{\text{H}}{\diagdown}}\text{C}=\text{C}\underset{\text{Cl}}{\overset{\text{H}}{\diagup}}$$

Cis isomer

$$\underset{\text{H}}{\overset{\text{Cl}}{\diagdown}}\text{C}=\text{C}\underset{\text{Cl}}{\overset{\text{H}}{\diagup}}$$

Trans isomer

If the two groups attached to either of the carbon atoms of the double bond are the same, cis-trans isomers aren't possible. Cis isomers are the normal form of fatty acids, whereas food processing tends to convert some of the cis isomers to the trans isomers.

Cis-trans isomers are also possible in cyclic systems. The cis form has similar groups on the same side of the ring, whereas the trans form has similar groups above and below the ring.

Chiral carbons

Trying to put your gloves on the wrong hands is kind of like another property of biological systems: handedness. *Left-handed* molecules rotate plane-polarized light to the left, and *right-handed* molecules rotate plane-polarized light to the right.

Identifying chiral molecules

The presence of an asymmetric, or *chiral,* carbon atom is sufficient to produce a "handed" molecule.

A chiral carbon atom has four different groups attached to it. The majority of biological molecules have one or more chiral carbon atoms and, for this reason, they're chiral. Figure 3-8 shows the chiral nature of glucose.

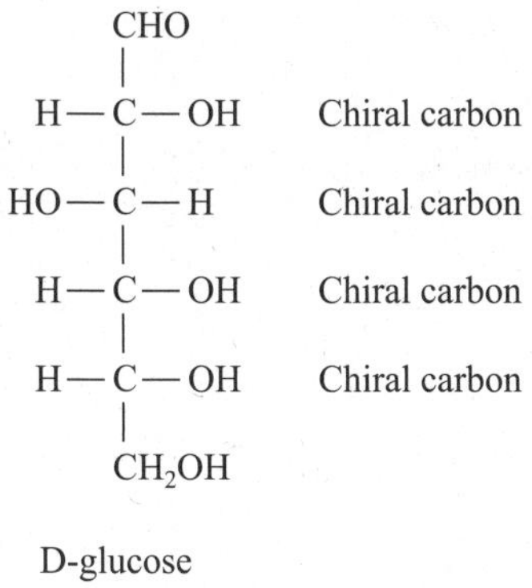

```
                    CHO
                     |
              H — C — OH          Chiral carbon
                     |
           HO — C — H             Chiral carbon
                     |
              H — C — OH          Chiral carbon
                     |
              H — C — OH          Chiral carbon
                     |
                   CH₂OH

              D-glucose
```

Figure 3-8: The structure of glucose, a sugar with four chiral carbon atoms.

Determining the chiral form: Enantiomer or sterioisomer?

All substances have a mirror image (okay, except vampires); however, if a chiral carbon atom is present, the mirror images are nonsuperimposable. Hold out your left and right hands, palms up — they are nonsuperimposable mirror images. These two mirror images are called *enantiomers.* The different chiral forms differ from each other in two aspects:

- ✔ How they affect light
- ✔ How they interact with other chiral substances (usually only one chiral form is biologically active)

To determine how a particular form affects light, it's necessary to use *plane polarized light,* in which all the light waves vibrate in the same plane. When you use this kind of light, a chiral substance rotates the vibrational plane of the light — one form (the dextrorotatory, *d*, (+) isomer) rotates the plane to the right, and the other (the levorotatory, *l*, (–) isomer) rotates the plane to the left. The *d* and *l* forms are *stereoisomers* and are optically active.

Illustrating the chiral compound: Fischer projection formulas

A chemist named Emil Fischer developed a method of drawing a compound to illustrate which stereoisomer was present. These Fischer projection formulas are very useful in biochemistry. In a projection formula, a chiral carbon is placed in the center of a + pattern. The vertical lines (bonds) point away from the viewer, and the horizontal lines point toward the viewer (see Figure 3-9). Fischer used the *D* designation if the most important group (the group whose central atom had the largest atomic number) was to the right of the carbon, and the *L* designation if the most important group (lowest atomic number) was to the left of the carbon. Figure 3-10 shows two Fischer projection formulas.

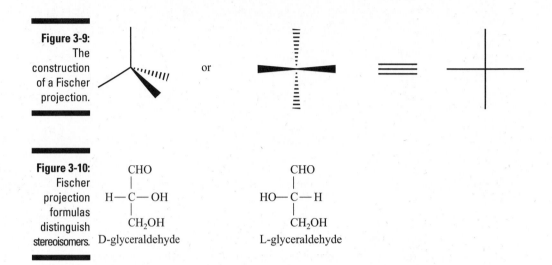

Figure 3-9: The construction of a Fischer projection.

Figure 3-10: Fischer projection formulas distinguish stereoisomers.

The *d* and *l* symbols aren't necessarily the D and L forms, respectively; thus, confusion may occur and lead to incorrect predictions. For this reason, the use of *d* and *l* is diminishing. The use of D and L is gradually being replaced by the R and S system of designating isomers. This system is particularly useful when more than one chiral carbon atom is present. For a description of this system, see *Organic Chemistry For Dummies* by Arthur Winter (Wiley).

Wow, that's all the important organic chemistry you need for biochemistry in about 16 pages. How many pages did your organic chemistry text have, and what did it cost? This book is a real bargain!

Part II
The Meat of Biochemistry: Proteins

The 5th Wave   By Rich Tennant

"Who wants to help Grandma make her famous gingerbread man cookies? You kids get the flour, eggs, and sugar, and I'll get the amino acids and enzymes."

In this part . . .

We focus, not surprisingly, on proteins, starting with amino acids, protein's building blocks. After that we detail the processes of amino acid sequencing and the various kinds of protein structure. We finish up this part by discussing enzyme kinetics, covering catalysts (which speed up reactions) and inhibitors (which slow them down).

Chapter 4

Amino Acids: The Building Blocks of Protein

*A*ll cells contain thousands of types of proteins, and amino acids are the building blocks of those proteins. The sequential order, number, and chemical identity of the amino acids in the protein determine the protein's structure and functions. That's why you need to understand the chemical properties of amino acids before you can understand the behavior of proteins.

Amino acids are relatively simple molecules containing both an amine group and an acid group. The biologically important amino acids are the α-amino acids, which have the amine and acid groups attached to the same carbon atom. More than 100 natural amino acids exist, but only 20 of them are used in protein synthesis. Francis Crick (who, with James Watson, determined the structure of DNA) labeled this set of amino acids the *magic 20.* (Sounds like the cast of a Harry Potter movie.) Other amino acids are found in certain proteins, but in almost all cases, these additional amino acids result from the modification of one of the magic 20 after the protein formed.

In this chapter, we examine the structure and properties of amino acids, especially the more common ones, and we show how they interact and combine.

General Properties of Amino Acids

In any organic compound, the functional groups that are present largely determine the molecules' properties. In biological systems, the important amino acid properties include the following:

✔ **They can join to form proteins.** The average molecular weight of an amino acid is about 135. Proteins have molecular weights ranging from about 6,000 to several million. Thus, a large number of amino acids must be joined together to produce a protein.

✔ **They all have both an acid and a base.** The α-carbon (end carbon) not only has an amine group ($-NH_2$) and a carboxylic acid group ($-COOH$) but also has two additional groups: a hydrogen atom and an R– group. The side chain, R group, identifies the amino acid.

✔ **They all have variations in what part of the structure is protonated, depending on the solution's pH and the rest of the molecule's structure.**

✔ **They all, except glycine, have a *chiral* nature (four different groups attached to the same carbon), influencing the reactions that the compound undergoes.**

Amino acids are positive and negative: The zwitterion formation

The presence of both an acid and a base (amine) in the same molecule leads to an interaction between the two that results in the transfer of a hydrogen ion from the acid portion to the base portion. An amino acid with both positive and negative regions is called a *zwitterion*. The net charge of the zwitterion as a whole is 0. The acid end of the amino acid has a negative charge ($-COO^-$), and the base end has a positive charge ($-NH_3^+$). The *deprotonated* portion (portion that has lost a hydrogen ion) is a carboxylate group, and the *protonated* portion (portion that has gained a hydrogen ion) is an ammonium group. The presence of a charge on the amino acid makes the amino acid water-soluble. Figure 4-1 shows a zwitterion's formation.

Figure 4-1:
A zwitterion's formation.

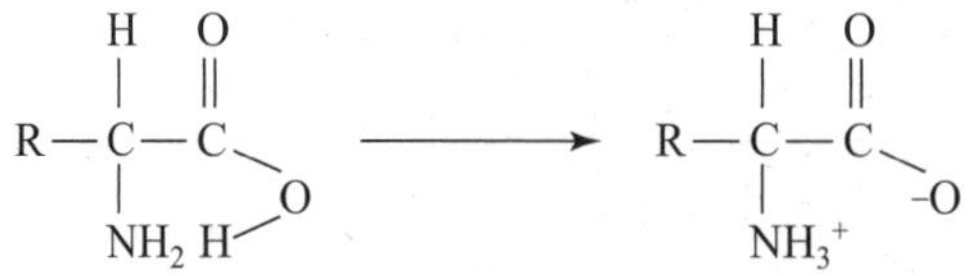

The unionized (*un-ionized,* not *unionized* — no Teamsters here) amino acid molecule shown in Figure 4-1 doesn't actually exist. However, many books and instructors draw the unionized form as a simplification, as if the ionization didn't occur.

Protonated? pH and the isoelectric point

Because of their acid-base nature, how amino acids react depends on the pH of the solution in which they're found. In this section, we look at some of the implications of this pH dependency. The zwitterion is the predominant form at a particular pH, which is designated the *isoelectric point* (pI). The isoelectric point is midway between the two different pK_a values. Under most physiological conditions, isolated amino acids exist in their zwitterion form (see Figure 4-2 [a]). Pure amino acids are also in the zwitterion form — and, for this reason, are *ionic solids.*

✔ **At a pH below the isoelectric point, some of the carboxylate groups are protonated** (see Figure 4-2 [b]). The pH required to cause this protonation depends on the acid's K_a. For this reason, the pK_a of the carboxylic acid group is important. Typical values are between 1 and 3. If, for example, the pK_a is 2.5, at a pH of 2.5, 50 percent of the carboxylate groups are protonated. The net charge of the protonated form is +1.

✔ **At a pH above the isoelectric point, some of the ammonium groups are deprotonated** (see Figure 4-2 [c]). The pH required to cause this deprotonation depends on the ammonium group's K_a. For this reason, the pK_a of the ammonium group is important. Typical values are between 8 and 11. If, for example, the pK_a is 10, at a pH of 10, 50 percent of the ammonium groups are deprotonated. The net charge of the deprotonated form is –1.

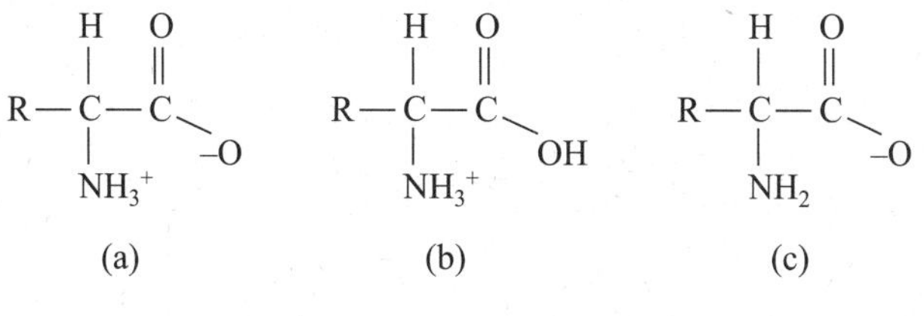

Figure 4-2:
(a) Zwitterion form, (b) protonated form, and (c) deprotonated form.

$$
\begin{array}{ccc}
\underset{\text{(a)}}{\overset{\displaystyle H\quad O}{R-\overset{|}{\underset{|}{C}}-\overset{\|}{C}\diagdown_{-O}}}\ \ &
\underset{\text{(b)}}{\overset{\displaystyle H\quad O}{R-\overset{|}{\underset{|}{C}}-\overset{\|}{C}\diagdown_{OH}}}\ \ &
\underset{\text{(c)}}{\overset{\displaystyle H\quad O}{R-\overset{|}{\underset{|}{C}}-\overset{\|}{C}\diagdown_{-O}}}
\end{array}
$$

(a) NH_3^+ (b) NH_3^+ (c) NH_2

Some of the side chains are also acidic or basic. In these cases, an additional pK_a becomes significant in the reactions of these molecules and obviously complicates the amino acid's pH behavior.

Asymmetry: Chiral amino acids

In a typical α-amino acid, four different groups are attached to the α-carbon (–COOH, –NH$_2$, –R, and –H). This makes the α-carbon *asymmetric* or *chiral*. The only exception is the amino acid glycine, where the R– group is a hydrogen atom. The presence of two hydrogen atoms on the α-carbon means that, as in the case of glycine, the carbon atom is *achiral*. Chiral materials are optically active; the different forms affect light in different ways. (See Chapter 3 for more on what makes a molecule chiral.)

The arrangement of the groups around a chiral carbon atom is important. Just as your left hand only fits into your left glove, only certain arrangements of the groups fit around a chiral carbon atom (because of what's called *handedness*). (That's why mittens are popular with parents of young children — they don't have handedness.)

Chiral amino acids come in two different forms: the D- and the L- forms. Only the L- forms are constituents of proteins. The D- forms appear in some antibiotics and in the cell walls of certain bacteria. Fischer projections, as we explain in Chapter 3, are commonly used to represent the arrangement about the chiral carbon. Figure 4-3 illustrates some different ways to draw the Fischer projections of the structure of amino acids.

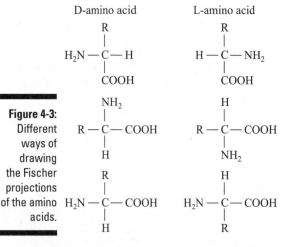

Figure 4-3: Different ways of drawing the Fischer projections of the amino acids.

A few amino acids contain two asymmetric carbon atoms. These cases have four possible isomers (see Chapter 3 for a discussion of isomers). Biological activity is usually limited to only one of these four isomers.

The Magic 20 Amino Acids

Amino acids are divided into four subgroups based on the nature of the side chain (groups attached to the α-carbon) and the general behavior of the amino acid. Those subgroups are

- ✔ Nonpolar (hydrophobic) and uncharged
- ✔ Polar (hydrophilic) and uncharged
- ✔ Acidic (polar and charged)
- ✔ Basic (polar and charged)

The properties of the side chains are important not only to the behavior of the individual amino acids but also to the properties of the proteins resulting from the combination of certain amino acids.

In the following sections we show you the structures of the individual amino acids. You can represent each of the amino acids by either a three-letter or a one-letter abbreviation. Like the chemical symbols for the elements, these are fixed abbreviations. The three-letter abbreviations are easier to relate to the name of the specific amino acid. For example, we use *gln* to represent glutamine. The one-letter abbreviations are shorter but not always related to the name. For example, we use *Q* for glutamine.

Nonpolar (hydrophobic) and uncharged amino acids

The nonpolar amino acids are as follows:

- ✔ Alanine (Ala, A)
- ✔ Isoleucine (Ile, I)
- ✔ Leucine (Leu, L)
- ✔ Methionine (Met, M)
- ✔ Phenylalanine (Phe, F)
- ✔ Proline (Pro, P)
- ✔ Tryptophan (Trp, W)
- ✔ Valine (Val, V)

Figure 4-4 shows these amino acids.

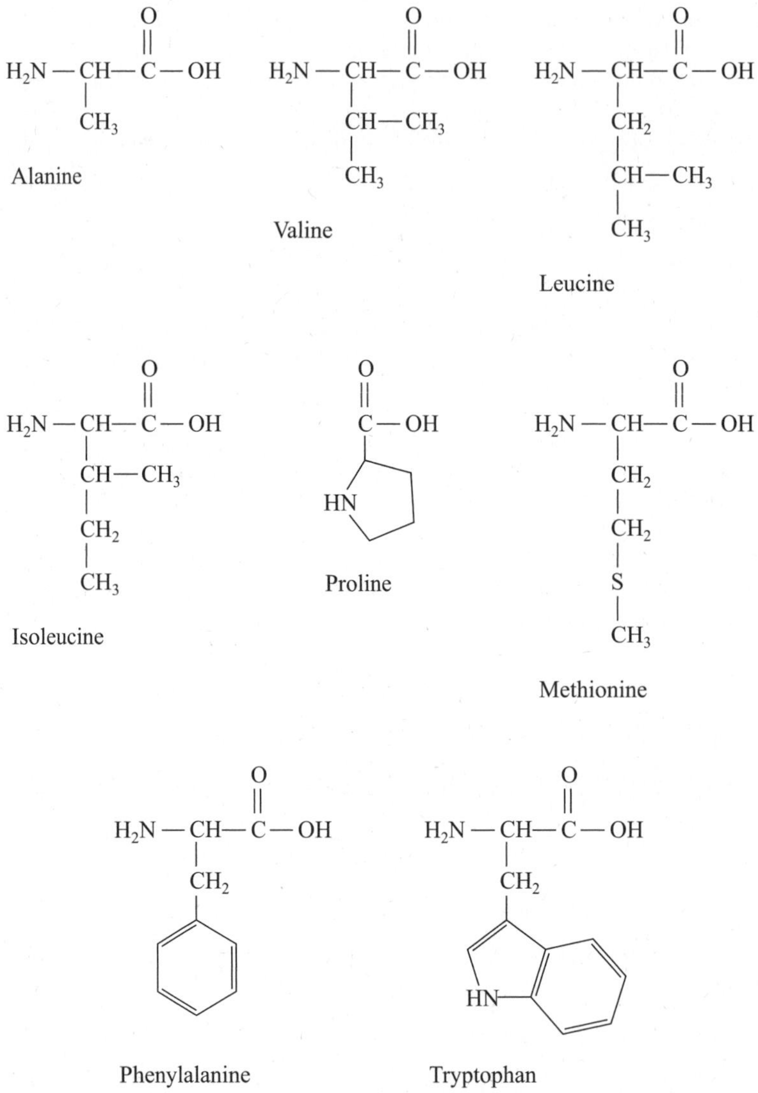

Figure 4-4:
Nonpolar
amino acids.

Proline has an unusual cyclic structure, which has a significant influence on protein structure. Tryptophan is a borderline case because the –NH from the ring system can interact with water to a limited extent.

Drawing the structures of amino acids

The simplified representations of the amino acids in Figure 4-4 don't actually occur in biological systems (or nonbiological systems). The amino acids always occur as ions. They may occur as *zwitterions* (have both positive and negative charges), *cations* (at low pH), or *anions* (at high pH). At low or high pH values, one end of the amino acid retains its charge and the other becomes neutral. The figure shown here illustrates the various ions that occur for all amino acids. The pH values necessary to convert from the zwitterion to one of the other forms vary from one amino acid to another.

Many of the amino acids have side chains (R groups) that are also acidic or basic. The form (cationic, anionic, or neutral) of these groups is also pH-dependent.

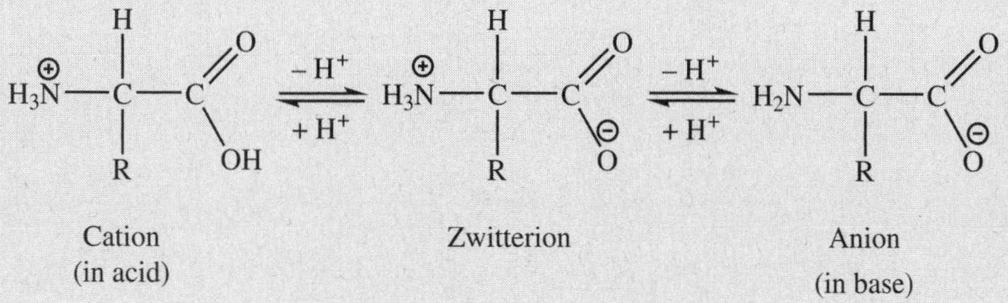

Cation
(in acid)

Zwitterion

Anion
(in base)

Polar (hydrophilic) and uncharged amino acids

The polar and uncharged amino acids, other than glycine, can hydrogen bond to water. For this reason, they're usually more soluble than the nonpolar amino acids. The amino acids in this group are as follows:

- ✔ Asparagine (Asn, N)
- ✔ Cysteine (Cys, C)
- ✔ Glutamine (Gln, Q)
- ✔ Glycine (Gly, G)
- ✔ Serine (Ser, S)
- ✔ Threonine (Thr, T)
- ✔ Tyrosine (Tyr, Y)

Glycine seems to be an unexpected member of this group. The small size of the R group in glycine leads to the predominance of the amino and carboxylate functional groups, resulting in glycine's similarity to other amino acids in this group. The amide, alcohol, and sulfhydryl (–SH) groups of the remaining members of this group are very polar and neutral. At very high pH values, the phenolic group on tyrosine ionizes to yield a polar charged group. Figure 4-5 shows these amino acids.

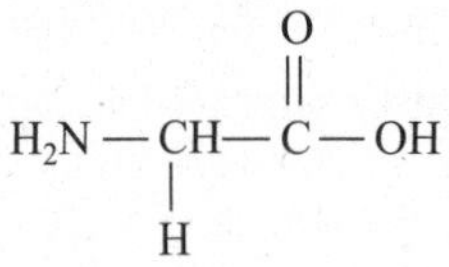

Glycine

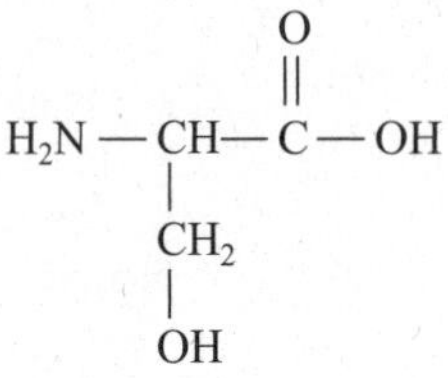

Serine

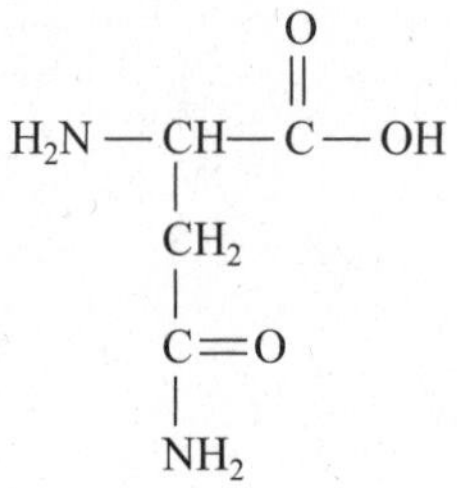

Asparagine

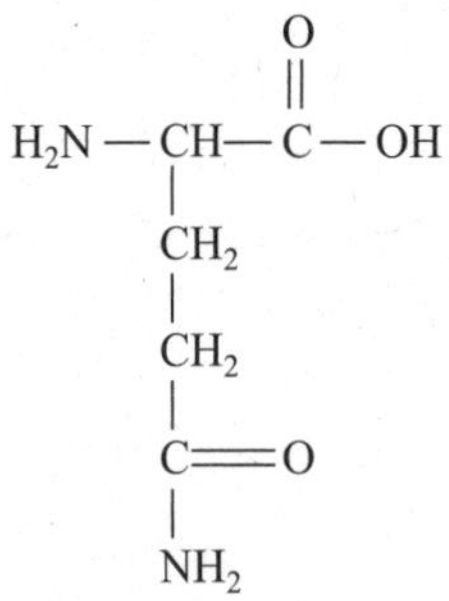

Glutamine

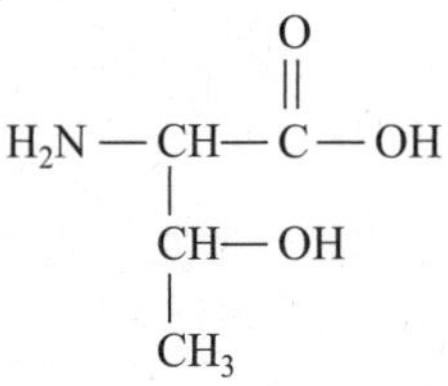

Threonine

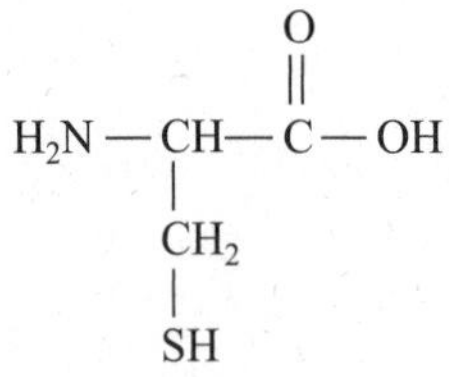

Cysteine

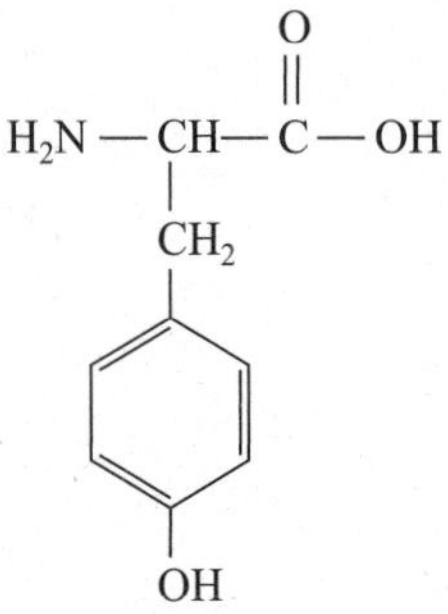

Tyrosine

Figure 4-5: Polar amino acids.

Acidic amino acids

The acidic amino acids are as follows:

- ✔ Aspartic acid (Asp, D)
- ✔ Glutamic acid (Glu, E)

In both of these amino acids, the side group contains a carboxylic acid group. This secondary carboxylic acid group is a weaker acid (higher pK_a) than the primary carboxylic acid group. This additional carboxylate group leads to a net –1 charge at a pH where the "normal" zwitterion has a 0 net charge. The carboxylate side chain is important in the interaction of many proteins with metal ions, as *nucleophiles* (an electron-rich group replacing some group attached to a carbon) in many enzymes, and in ionic interactions. Figure 4-6 shows these amino acids.

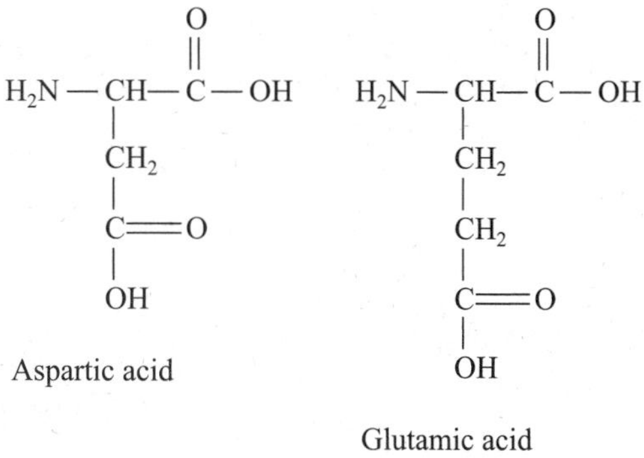

$$
\begin{array}{cc}
\underset{\displaystyle \text{Aspartic acid}}{
\begin{array}{l}
\quad\quad\quad \overset{\displaystyle O}{\underset{\displaystyle \|}{}} \\
H_2N-CH-C-OH \\
\quad\quad | \\
\quad\quad CH_2 \\
\quad\quad | \\
\quad\quad C{=}O \\
\quad\quad | \\
\quad\quad OH
\end{array}}
&
\underset{\displaystyle \text{Glutamic acid}}{
\begin{array}{l}
\quad\quad\quad \overset{\displaystyle O}{\underset{\displaystyle \|}{}} \\
H_2N-CH-C-OH \\
\quad\quad | \\
\quad\quad CH_2 \\
\quad\quad | \\
\quad\quad CH_2 \\
\quad\quad | \\
\quad\quad C{=}O \\
\quad\quad | \\
\quad\quad OH
\end{array}}
\end{array}
$$

Figure 4-6: Acidic amino acids.

Basic amino acids

The basic amino acids are as follows:

- ✔ Arginine (Arg, R)
- ✔ Histidine (His, H)
- ✔ Lysine (Lys, K)

All these are classified as basic amino acids, but dramatic changes in pH can affect their reactivities. This is especially true of histidine.

All three of these amino acids have a basic group that's capable of accepting a hydrogen ion. In the case of lysine, this is a simple ammonium ion.

Arginine forms the guanidinium group. Histidine forms an imidazolium group. (We don't know who came up with this naming system!) As in the case of the acidic side chains, these side chains have a pK_a value. Both arginine and lysine side chains are usually protonated at physiological pH values. In these cases, there's a net +1 charge present. In proteins, this net charge may be part of an ionic interaction. The pK_a of the side chain of histidine is lower than other basic groups. Protonation of histidine becomes significant at much lower pH values. In many proteins, histidine is not protonated but is important in many enzymes in the hydrogen ion transfer processes. Figure 4-7 shows these basic amino acids.

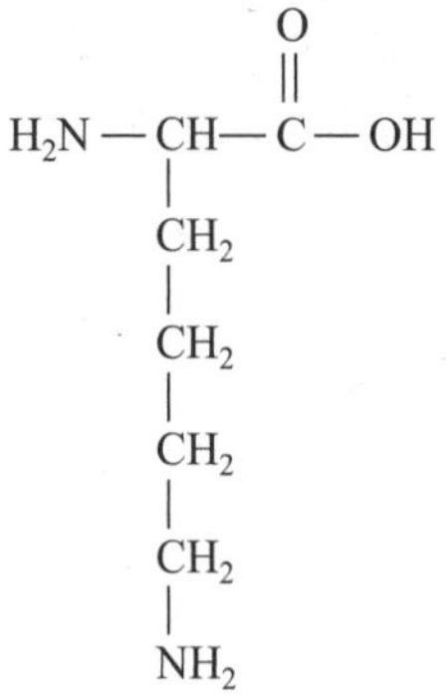

Lysine

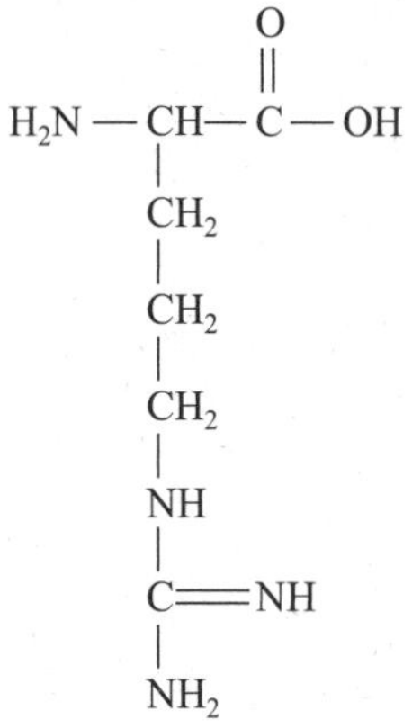

Arginine

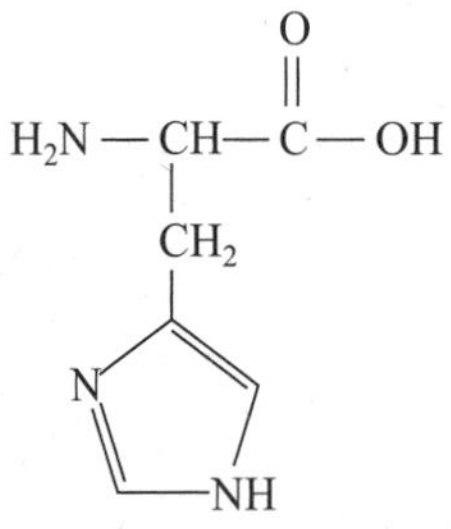

Histidine

Figure 4-7:
Basic amino acids.

Lest We Forget: Rarer Amino Acids

In a few cases, an amino acid may undergo modification after it's incorporated into a protein. Collagen and gelatin, for example — proteins present in higher vertebrates — contain hydroxylysine and hydroxyproline. These two amino acids contain an additional –OH group on the side chain.

Certain amino acids don't occur in proteins. The neurotransmitter *γ-aminobutyric acid* — GABA — is one example. *Citrulline* is the amino acid that serves as a *precursor* (substance used in the creation) of arginine. *Ornithine, homocysteine,* and *homoserine* are important as metabolic intermediates. Figure 4-8 shows a couple of these amino acids.

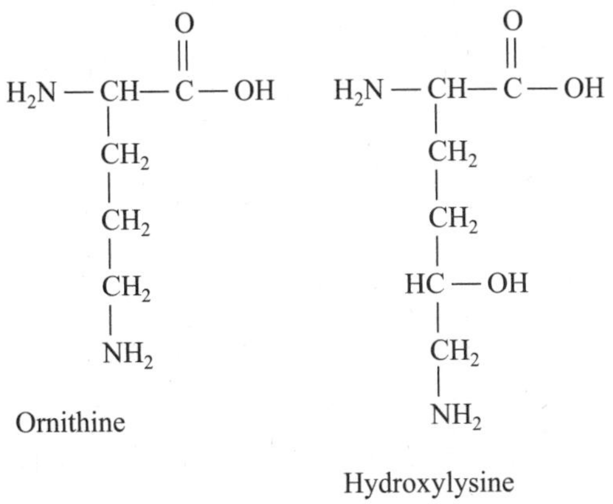

$$H_2N - CH - C - OH$$

Figure 4-8:
Two of the less common amino acids.

Ornithine

Hydroxylysine

Rudiments of Amino Acid Interactions

Amino acids are the ingredients used in the recipe to make a protein. Just as the individual ingredients in a recipe lead to distinct characteristics of what eventually shows up on the dinner table, the amino acids contribute properties to proteins. And just as you can't replace the flour in a recipe with pepper, you generally can't replace one amino acid in a protein with another. In both cases, the final product will be different. In this section, we show you some of the ways that amino acids interact. These interactions set the stage for our discussion of bonding among the amino acids in the section "Combining Amino Acids: How It Works," later in the chapter.

Intermolecular forces: How an amino acid interacts with other molecules

Amino acids can interact with other molecules — and we mean *any* other molecules, including fluids, other amino acids, and other biological molecules — in a variety of ways. We cover intermolecular forces in general in Chapter 3, but in this section we show you how intermolecular forces play out when amino acids are involved. The carboxylic acid and amine parts of the amino acids define much of the molecule's reactivity, but the side chains can also interact with other molecules. Here are three general ways in which the side chains can interact:

✔ **Hydrophobic interactions:** The nonpolar side groups are hydrophobic and are attracted to each other through *London dispersion forces* (see Chapter 3 for a brief review of London dispersion forces). Nonpolar groups tend to clump together and exclude not only water but also all other types of side chains.

✔ **Hydrophilic reactions:** The polar and uncharged side groups are hydrophilic. The presence of a number of these groups increases a protein's solubility. These groups hydrogen bond not only to water but also to each other. Polar groups tend to interact strongly and "push" the nonpolar groups out.

✔ **Ionic interactions:** The presence of acidic or basic side chains leads to ionic charges — opposite charges attract. A carboxylate group from one side chain is attracted to the ammonium ion of another side chain through an ionic interaction. This ionic bond is very strong.

The amino acid cysteine can interact with a second cysteine molecule through a different type of interaction (see Figure 4-9). The mild oxidation of two cysteine sulfhydryl groups leads to the formation of cystine. A disulfide linkage joins the two amino acids with a covalent bond. Mild reduction can reverse this process.

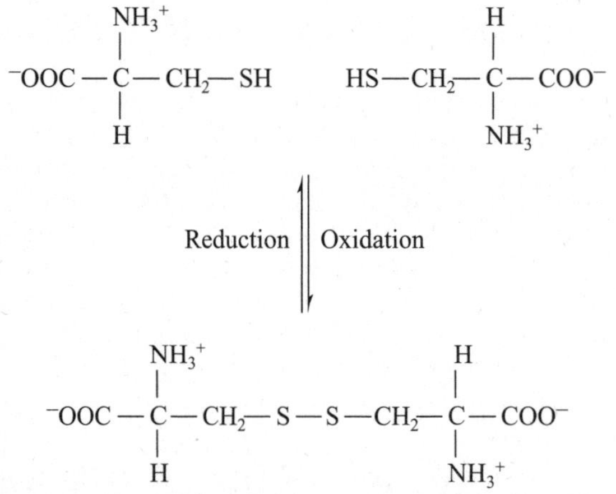

$$^-OOC-\underset{\underset{H}{|}}{\overset{\overset{NH_3^+}{|}}{C}}-CH_2-SH \qquad HS-CH_2-\underset{\underset{NH_3^+}{|}}{\overset{\overset{H}{|}}{C}}-COO^-$$

Reduction ‖ Oxidation

Figure 4-9:
Joining two cysteine molecules to form cystine.

$$^-OOC-\underset{\underset{H}{|}}{\overset{\overset{NH_3^+}{|}}{C}}-CH_2-S-S-CH_2-\underset{\underset{NH_3^+}{|}}{\overset{\overset{H}{|}}{C}}-COO^-$$

A hair perm uses an oxidation reduction reaction, creating disulfide linkages. The greater the number of disulfide linkages, the curlier the hair! And the smell . . . explained!

Altering interactions by changing the pH

As we discuss in Chapter 3, the function of many substances, especially bio-chemical ones, is dependent on pH. In most biochemical cases, an oxygen, nitrogen, sulfur, or phosphorus molecule is involved. If you change the pH, you change some of the interactions. In this section we show how those changes affect interactions involving amino acids.

Just like many other molecules, an amino acid has two or three functional groups, depending on the amino acid. These functional groups include those with oxygen and sulfur and those with nitrogen. A change in pH affects one to three of these functional groups in terms of interactions. So if an amino acid has a functional group that changes from a dipole-dipole interaction to an ionic interaction, the properties of that amino acid change.

One example of the dipole-dipole to ionic interaction change is the process of milk curdling. If you add an acid to milk, it coagulates. Casein has an isoelec-tric point at a pH of 4.6, so adding an acid causes the formation of ionic bonds among the molecules. This works against the dipole-dipole interactions with water so that the protein precipitates.

Table 4-1 shows the pK_a values for the various groups present in the different amino acids. If the pH of the solution matches one of these values, then half the species is in the protonated form and half is in the deprotonated form. At a lower pH, more than half is protonated, whereas at a higher pH, more than half is deprotonated.

The pH dependence of the protonation of amino acids aids in their separation and identification. Because the amino acids use the carboxylic acid and amine ends when they join to form a protein, only the pK_a values of the side chains are important in additional interactions and reactions.

Table 4-1	pK_a Values for the Amino Acids		
Amino Acid	*pK_a –COOH*	*pK_a –NH_3^+*	*pK_a R Group*
Alanine	2.35	9.69	
Arginine	2.17	9.04	12.48
Asparagine	2.02	8.80	
Aspartic acid	2.09	9.82	3.86

(continued)

Table 4-1 (continued)

Amino Acid	pK$_a$ –COOH	pK$_a$ –NH$_3^+$	pK$_a$ R Group
Cysteine	1.71	10.78	8.33
Glutamic acid	2.19	9.67	4.25
Glutamine	2.17	9.13	
Glycine	2.34	9.60	
Histidine	1.82	9.17	6.00
Isoleucine	2.36	9.68	
Leucine	2.36	9.60	
Lysine	2.18	8.95	10.53
Methionine	2.28	9.21	
Phenylalanine	1.83	9.13	
Proline	1.99	10.60	
Serine	2.21	9.15	
Threonine	2.63	10.43	
Tryptophan	2.38	9.39	
Tyrosine	2.20	9.11	10.07
Valine	2.32	9.62	

Combining Amino Acids: How It Works

A *protein* is a string of at least 150 amino acids (residues) joined together. We cover the fundamentals of protein creation in Chapter 5, but before you dive into that topic, this section gives you a solid understanding of how two amino acids join together in the first place and how additional amino acids link onto the chain gang. The process is reversible (as in digestion).

When drawing the chemical structures of amino acids and their bonds, the standard convention is to first draw the structures from the ammonium group of the first amino acid (the N-terminal residue), starting at the left, and to continue drawing to the right, ending with the carboxylate group (C-terminal residue) of the last amino acid.

The peptide bond and the dipeptide

One of the most important types of bonds in all of biochemistry is the *peptide bond*. This type of bond is used in the synthesis of proteins. The interaction of two amino acids at the body's pH results in the formation of a peptide bond, as illustrated in Figure 4-10.

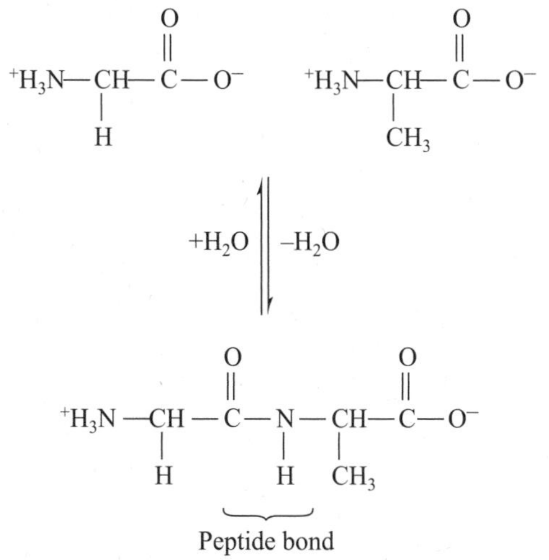

Figure 4-10: The formation of a peptide bond.

$$^+H_3N-CH-C-O^- \qquad ^+H_3N-CH-C-O^-$$

$$+H_2O \quad\|\quad -H_2O$$

$$^+H_3N-CH-C-N-CH-C-O^-$$

Peptide bond

The two residues react to expel a water molecule, the same dehydration reaction you use so much in organic chemistry. The reverse of this condensation reaction is hydrolysis. The resultant amide group is a peptide bond. The presence of two amino acid residues means the product is a dipeptide.

The peptide bond is a *flat* (planar) structure. It's stabilized by our old organic friend, *resonance.* Figure 4-11 illustrates the stabilization. The resonance increases the polarity of the nitrogen and oxygen. This increase in polarity leads to hydrogen bonds that are much stronger than most other hydrogen bonds. The double bond character between the carbon and the nitrogen restricts rotation about this bond. That's why the peptide bond is planar.

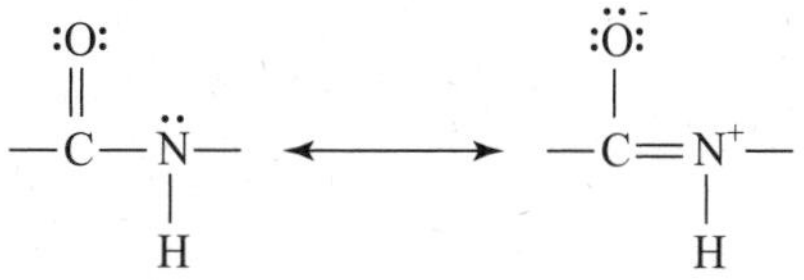

Figure 4-11: Resonance stabilization of a peptide bond.

$$-C-\ddot{N}- \quad\longleftrightarrow\quad -C=N^+-$$

Tripeptide: Adding an amino acid to a dipeptide

A repetition of the process illustrated in Figure 4-10 joins a third amino acid to produce a *tripeptide*. For example, combining glycine, alanine, and serine yields the illustration in Figure 4-12. Notice that everything begins with the N-terminal residue and ends with the C-terminal residue. (You could designate this tripeptide as *Gly-Ala-Ser* using the three-letter abbreviations.)

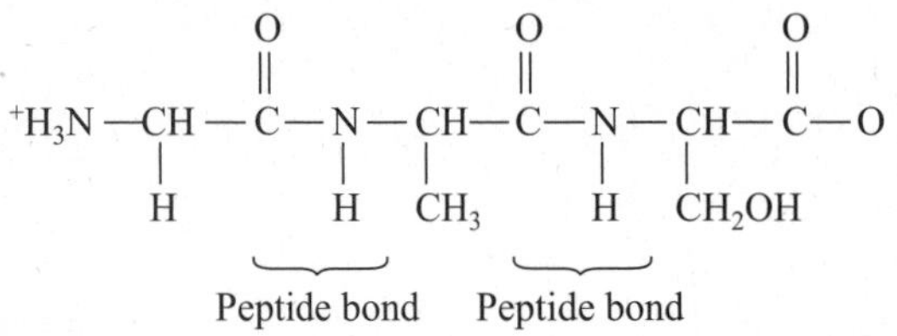

Figure 4-12: A tripeptide.

The repetition of the process of linking amino acids hundreds or thousands of times produces a protein. In the next chapter, we cover that topic in more detail.

Chapter 5

Protein Structure and Function

*I*n Chapter 4, we show you how amino acids combine through peptide bonds, and we note that if at least 150 or so amino acids join hands, they rise to the rank of a protein. However, distinguishing an amino acid chain as a protein isn't exactly simple. Just as written English is an extremely diverse set of words made by combining letters from an alphabet of just 26 letters, proteins are an extremely diverse set of biochemicals made by combining 20 different amino acids.

In this chapter, we discuss proteins in detail, including the four types of protein structure that determine a protein's function and the sequence of amino acids in a particular protein.

Proteins: Not Just for Dinner

Proteins fall into two general categories:

- ✔ **Fibrous proteins** are found only in animals. They usually serve as structural entities such as connective tissue, tendons, and muscle fiber. They're normally insoluble in water.

- ✔ **Globular proteins** don't usually serve a structural function. They can act as transporters, like hemoglobin, and are often enzymes. They're usually water-soluble.

Living organisms use proteins in a number of ways:

- ✔ **Structure:** Skin and bone contain *collagen,* a fibrous protein.

- ✔ **Catalysis:** Proteins called *enzymes* allow reactions to occur in an organism under mild conditions and with great specificity.

- ✔ **Movement:** Proteins make up a large percentage of muscle fiber and help in the movement of various parts of your body.

- ✔ **Transport:** Proteins transport small molecules through an organism. *Hemoglobin,* the protein that transports oxygen to the cells, is a transport protein.

- ✔ **Hormones:** Hormones that happen to be proteins help regulate cell growth.

- ✔ **Protection:** Proteins called *antibodies* help rid the body of foreign harmful substances.

- ✔ **Storage:** Some proteins help store other substances in an organism. For example, iron is stored in the liver in a complex with the protein *ferritin.*

- ✔ **Regulation:** Proteins help mediate cell responses, such as the protein *rhodopsin,* found in the eye and involved in the vision process.

The function that a particular protein assumes is, in many cases, directly related to the protein's structure. Proteins may have as many as four levels of structure (key word being *levels,* not different structures), each of which places the components into a position where intermolecular forces can interact most advantageously. The levels, which we discuss in the following sections, are simply labeled *primary, secondary, tertiary,* and *quaternary.* Primary is the most fundamental level that all proteins have, and quaternary is the most specific level that only some proteins have. Intermolecular forces themselves are important to a protein's function, of course, but the arrangement of the molecules is even more significant.

If present, the secondary, tertiary, and quaternary structures of a protein may be destroyed in a number of ways:

- ✔ Heating (cooking) can break hydrogen bonds.

- ✔ Changing the pH can protonate or deprotonate the molecule and interrupt ionic interactions.

- ✔ Reducing agents can break disulfide linkages.

In some cases, the process may be reversible.

Now you know why I (coauthor John) consider myself a protein kind of guy. Not only do I love steak, chicken, and so on, but a large part of my own body is protein. So let's dive into that protein pool.

Primary Structure: The Structure Level All Proteins Have

The *primary structure* of a protein is simply the sequence of amino acids that comprise the molecule. All proteins have a primary structure because all proteins by definition consist of a sequence of amino acids. The primary structure serves as the foundation upon which all higher levels of protein structure build.

But how do those amino acids become connected in a specific order? Following is a look at how a protein is assembled from its building blocks, the amino acids.

Building a protein: Outlining the process

During the synthesis of a protein, the chain of amino acids is built one link at a time, roughly as follows:

1. **A transfer RNA (tRNA) molecule transfers specific amino acids to the ribosomes of the cell to connect to the growing chain.**

2. **Each amino acid joins to the chain through the formation of a peptide bond.** (See Chapter 4 for more on peptide bonds.)

3. **The first peptide bond joins two amino acids to form a dipeptide.**

4. **The second peptide bond joins three amino acids to produce a tripeptide.**

5. **This process continues for hundreds, if not thousands, of times to produce a polypeptide — a protein.**

When two or more amino acids combine, a molecule of water is removed. What remains of each amino acid is called a *residue,* which lacks a hydrogen atom on the amino group, an –OH on the carboxyl group, or both.

The cell's DNA ultimately controls the sequence of amino acids. This information goes from the DNA to the messenger RNA (mRNA), which serves as the template for the creation of the protein's primary structure. It's like a string of pearls, where each pearl represents an amino acid. However, this would be a *very* long necklace! In order for the protein to be synthesized, energy must be supplied.

Organizing the amino acids

One end of the primary structure has an amino group, and the other end has a carboxylate group. By convention, the amino group end is considered the "beginning" of the protein. When drawing, naming, or numbering the primary structure, you always start with the amino end (called the *N-terminal*) and finish with the carboxylate end (the *C-terminal*). For example, in the hexapeptide Met-Thr-Ser-Val-Asp-Lys (see Chapter 4 for a list of the amino acids and their abbreviations), methionine (Met) is the N-terminal amino acid, and lysine (Lys) is the C-terminal amino acid. Note that reversing the sequence to Lys-Asp-Val-Ser-Thr-Met results in a hexapeptide with the same composition but with different chemical properties because it starts with a different amino acid. Therefore, an amino acid that loses a hydrogen atom in one sequence loses an –OH in the other.

The polypeptide chain has a backbone that consists of the same, rather simple, repeating unit. You can see this repeating sequence in Figure 5-1. Variations take place in the form of side chains — the R groups of the amino acids. Notice that the repeating unit (indicated by the brackets) is the amino-carbon-carbonyl sequence and that different R groups can attach to the carbon unit.

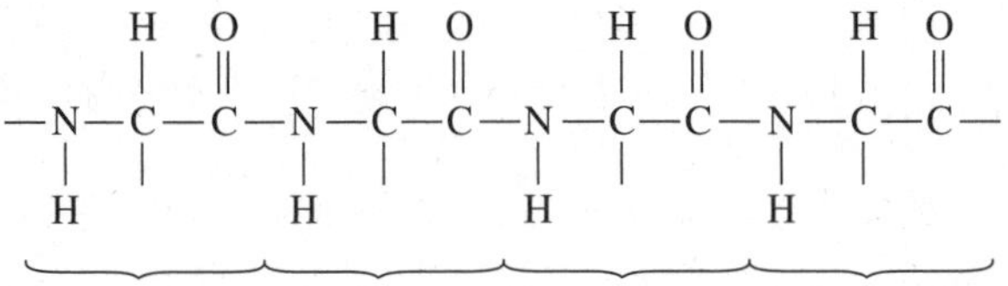

Figure 5-1:
Repeating sequence of the protein backbone.

The protein backbone has many places where hydrogen bonds may form. Every residue — other than the amino acid proline — has an NH, which may serve as a *donor* of a hydrogen bond. And every residue has a carbonyl group, which may serve as the *acceptor* of a hydrogen bond. The presence of donors and acceptors leads to the possibility of forming numerous hydrogen bonds. (We hope that you're gaining an appreciation of how very important intermolecular forces, especially hydrogen bonds, really are.)

The peptide bonds don't freely rotate about the carbon-nitrogen bond because of the contribution of the resonance form, which has a double bond. Thus, the backbone has a planar unit of four atoms, and in almost all cases, the oxygen atom is trans to the hydrogen atom. The remainder of the backbone can rotate. The ability to rotate influences how the protein's three-dimensional structure is established. This rotation is restricted because the side chains can "bump" into each other — an effect called *steric hindrance.* The peptide bond's rigidity and the rotation restrictions lower the entropy of the three-dimensional structure of a protein relative to a random chain of amino acids. Lowering the entropy helps stabilize the structure and obeys the laws of thermodynamics.

Example: The primary structure of insulin

The first determination of the primary structure of a protein was that of bovine insulin, the structure of which appears in Figure 5-2. Since this landmark determination, the primary structures of more than 100,000 proteins have been determined. In all cases, the protein has a unique primary structure.

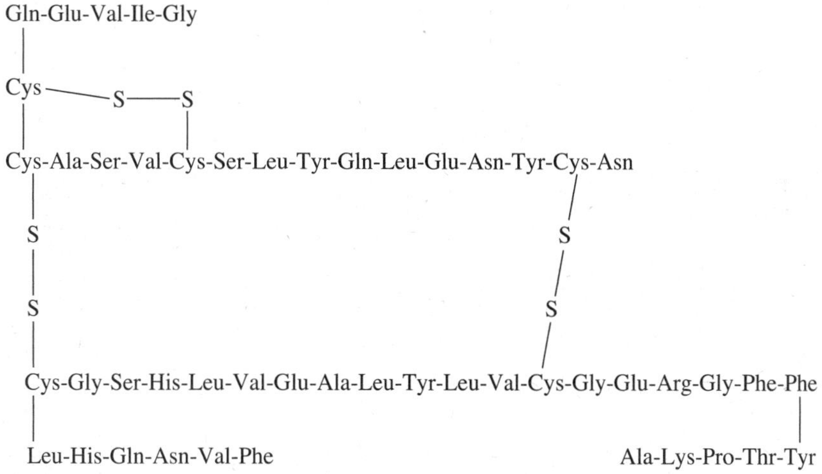

Gln-Glu-Val-Ile-Gly
|
Cys———S———S
| |
Cys-Ala-Ser-Val-Cys-Ser-Leu-Tyr-Gln-Leu-Glu-Asn-Tyr-Cys-Asn
| /
S   S
| /
S   S

Figure 5-2:
Structure
of bovine
insulin.

Cys-Gly-Ser-His-Leu-Val-Glu-Ala-Leu-Tyr-Leu-Val-Cys-Gly-Glu-Arg-Gly-Phe-Phe
| |
Leu-His-Gln-Asn-Val-Phe   Ala-Lys-Pro-Thr-Tyr

Secondary Structure: A Structure Level Most Proteins Have

Let's go back to our string of pearls analogy. The string can be folded back onto itself. This doesn't affect the order of pearls, but it gives the necklace a different look. The same thing can happen with our string of amino acids. Two peptide bonds can bond to each other through a hydrogen bond, as shown in Figure 5-3. In general, the formation of these hydrogen bonds leads to a protein's *secondary structure,* which is the result of many hydrogen bonds, not just one. The hydrogen bonds are *intramolecular* — that is, between segments of the same molecule.

Figure 5-3:
Hydrogen
bonding
between
two peptide
bonds.

$$-N-H\cdots\cdots O=C-$$
| |

The α-helix and β-pleated sheet are the secondary structures that result from this hydrogen bonding. Secondary structures may be only a small portion of a protein's structure or they can make up 75 percent or more.

The α-helix

In the α-helix (see Figure 5-4), the primary structure twists into a tightly wound, spring- or rod-like structure. Many times this is represented as a curly ribbon. Each turn consists of 3.6 amino acid residues. These turns allow hydrogen bonding between residues spaced four apart. Every peptide bond participates in two hydrogen bonds: one from an NH to a neighboring carbonyl, and one from a neighboring NH to the carbonyl.

Structurally, the helixes may be either right-handed or left-handed (see Chapter 3 for more on handedness). Essentially all known polypeptides are right-handed. Slightly more steric hindrance is present in a left-handed helix, and the additional steric hindrance makes the structure less stable. *Keratin* — the protein of fur, hair, and nails — consists of three right-handed α-helixes wrapped around one another in a left-handed coil.

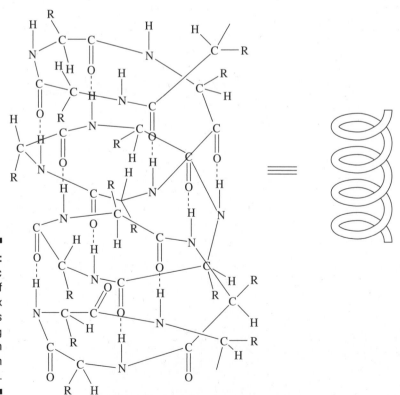

Figure 5-4:
The generic structure of an α-helix with its corresponding ribbon diagram representation.

Certain amino acids destabilize the α-helix. Proline, for example, creates bends or "kinks" in the primary structure that inhibit the formation of a regular pattern of hydrogen bonds. A group of isoleucine residues disrupts the secondary structure because of the steric hindrance caused by their bulky R groups. The small R group of glycine, only an H, allows too much freedom of movement, which leads to a destabilization of the helix. A concentration of aspartic acid and/or glutamic acid residues also destabilizes the structure because the negative charges on the side chains repel each other. Other residues that destabilize the helix, for similar reasons, are lysine, arginine, serine, and threonine.

The β-pleated sheet

The β-pleated sheet, or simply the β sheet, is the other major secondary protein structure. Here, the primary structure is extended instead of tightly winding into a helix. This structure has two forms, known as the *parallel* β-pleated sheet and the *antiparallel* β-pleated sheet. (Wow! Parallel and antiparallel — sounds like concepts from a sci-fi movie, doesn't it?)

Again, hydrogen bonds are the source of these structures. A β-pleated sheet forms when two or more strands link by hydrogen bonds. The strands are different parts of the same primary structure.

In the parallel structure, the adjacent polypeptide strands align along the same direction from the N-terminal end to the C-terminal end. In the antiparallel structure, the alignment is such that one strand goes from the N-terminal end to the C-terminal end, and the adjacent strand goes from the C-terminal end to the N-terminal end (see Figure 5-5).

In the β-pleated sheet structures, the side chains of adjacent amino acids point in opposite directions. The hydrogen bonding pattern in the parallel structure results in a more complicated structure. Here, the NH group of one residue links to a CO on the adjacent strand, whereas the CO of the first residue links to the NH on the adjacent strand that's two residues down the strand. In the antiparallel structure, the NH and CO groups of one residue link to the respective NH and CO groups of one residue on the adjacent strand.

Schematically, broad arrows denote the presence of β-pleated sheets. Arrows that point in the same direction indicate parallel structure, and arrows that point in opposite directions indicate antiparallel structure. The sheets are typically four or five strands wide, but ten or more strands are possible. The arrangements may be purely parallel, purely antiparallel, or mixed.

Parallel

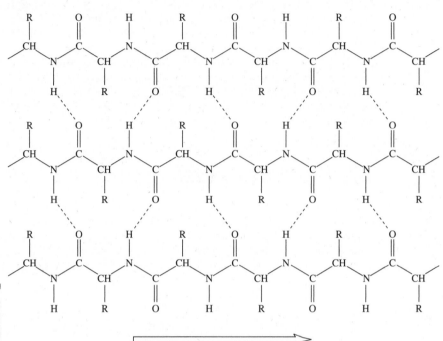

Figure 5-5:
Parallel and
antiparallel
β-pleated
sheet
structures.

β-turns and the Ω-loops

Additional secondary structures involve hydrogen bonding between peptide bonds; these are much smaller units. The best known are the β-turn — or *hairpin bend* — and the Ω-loop. The hairpin bend is simply a bend in the primary structure held in place by a hydrogen bond. The Ω-loop gets its name from the loose similarity of its shape to the Greek letter. Both are found on the exterior of proteins. Some proteins can have more than one type of secondary structure.

Antiparallel

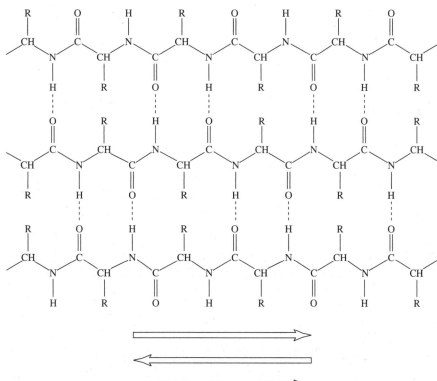

Antiparallel

Figure 5-5:
(continued)

Tertiary Structure: A Structure Level Many Proteins Have

A protein's primary and secondary structures, along with interactions between the side chains, determine the protein's overall shape. All of these primary and secondary structures along with the interactions give rise to what's called the protein's *tertiary structure*. (Let's take that folded string of pearls and put a couple of knots in it. It looks much different from the initial string.) Nonpolar side chains are hydrophobic and, although repelled by water, are attracted to each other. Polar side chains attract other polar side chains through either dipole-dipole forces or hydrogen bonds (see Figure 5-6).

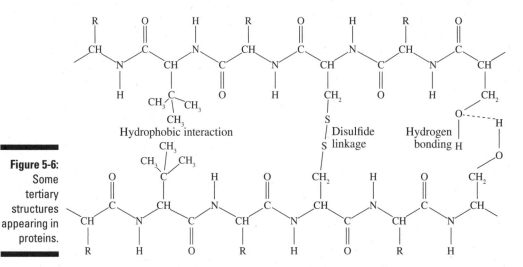

Figure 5-6:
Some
tertiary
structures
appearing in
proteins.

For example, both aspartic acid and glutamic acid have side chains with a negative charge that are strongly attracted to the positive charges in the side chains of lysine and arginine. Two cysteine residues can connect by forming a disulfide linkage — a covalent bond.

What induces a protein to adopt a very specific tertiary structure? Examination of the structures of many proteins shows a preponderance of nonpolar side chains in the interior with a large number of polar or ionic side chains on the exterior. In an aqueous environment, the hydrophobic (nonpolar) groups induce the protein to fold upon itself, burying the hydrophobic groups away from water and leaving the hydrophilic groups adjacent to water. The result is similar in structure to a micelle. This is especially important for structural and transport proteins.

Quaternary Structure: A Structure Level Some Proteins Have

The *quaternary structure* found in some proteins results from interactions between two or more polypeptide chains, interactions that are usually the same as those that give rise to the tertiary structure. These interactions include hydrogen bonding and disulfide bonds. This quaternary structure locks the complex of proteins into a specific geometry. An example is hemoglobin, which has four polypeptide chains — two identical α-chains and two identical β-chains. (The designations α and β simply refer to two different polypeptide chains [subunits] and not to secondary structures.)

Dissecting a Protein for Study

In the previous sections we discuss the different types of protein structure. Now we turn our attention to seeing how a biochemist goes about determining the structure(s) of a particular protein. (I bet a lot of budding young biochemists took their toys, and probably some of their parents' stuff, apart to see how they worked. The really good ones could put them back together.)

Additional information about a protein's structure comes from immunology. An animal generates an *antibody* — a protein found in the blood serum — in response to a foreign substance known as an *antigen*. Antigens collect on the surface of red blood cells. Exposure to bacteria or viruses, certain chemicals, and allergens induce the formation of certain antibodies. Every antigen has a specific antibody.

Antibodies have a strong affinity for their particular antigens, recognizing specific amino acid sequences on the antigens. This is like a key for a specific lock. Animals have a large number of antibodies present in their bodies, based on their environmental history. One application of antibodies and antigens is in the analysis of blood, specifically in the field of forensics investigations.

Separating proteins within a cell and purifying them

Each cell has thousands of different proteins. To examine and study one of them, you need to separate it from all the others. The methods of separating proteins are, in general, applicable to all other types of biochemicals. Initially, simple filtration and solubility can remove gross impurities, but you need to do much more before the sample is pure. The key separation and purification methods depend on two physical properties of the proteins: size and charge.

Separating proteins by size

Methods used to separate proteins by size and mass include ultrafiltration, ultracentrifugation, and size exclusion chromatography. *Ultrafiltration* is a modification of dialysis in which molecules smaller than a certain size diffuse through a semipermeable membrane and larger ones don't. Ultrafiltration can separate smaller molecules from larger impurities or larger molecules from smaller impurities.

In *ultracentrifugation,* a powerful centrifuge causes heavier molecules to sink faster, which allows their separation, much as the lighter water is separated from the heavier lettuce in a salad spinner. Ultracentrifugation can also be used to determine a protein's molar mass.

Forensics: Analysis of bloodstains

The study of proteins has many applications to forensics. One of them is the examination of bloodstains, blood being the most common form of evidence examined by a forensic serologist. The presence of blood can link a suspect to both a victim and a crime scene. Bloodstain patterns can also give evidence of how a violent attack took place. Criminals recognize the significance of this evidence and often try to conceal it.

Blood is mostly water, but it also contains a number of additional materials including cells, proteins, and enzymes. The fluid portion, or *plasma,* is mostly water. The *serum* is yellowish and contains platelets and white blood cells. The *platelets,* or red blood cells, outnumber the white blood cells by about 500 to 1. White blood cells are medically important, whereas red blood cells and, to a lesser extent, serum are important to the forensic serologist. Because blood quickly clots when exposed to air, serologists must separate the serum from the clotted material. The serum contains antibodies that have forensic applications, and red blood cells, in addition to DNA, have substances such as antigens on their surfaces that also are of interest to the forensic scientist. Antibodies and antigens are the keys to forensic serology: Even identical twins with identical DNA have different antibodies. As you know from this chapter, antibodies, and some antigens, are proteins, and that's why methods of studying proteins are important to the analysis of antibodies and antigens.

Analysis of bloodstains initially attempts to answer five questions.

✔ **Is this a blood sample?** To answer this question, the investigator can use a number of tests. The generic term for a test of this type is a *presumptive* test. The *Kastle-Meyer* test uses *phenolphthalein,* which,

when it comes into contact with hemoglobin or a few other substances, forms a bright pink color from the release of peroxidase enzymes. (We bet you remember seeing this test used in a number of TV shows.) The *luminol* test is useful in detecting invisible bloodstains because, in contact with blood or a few other chemicals, luminol emits light, which can be seen in a darkened room. The *Wagenhaar, Takayama,* and *Teichman* tests take advantage of the fact that long-dried blood crystallizes or can be induced to crystallize.

✔ **Is the blood from a human or an animal?** The forensic investigator answers this question (and the next one, if applicable) by means of an *antiserum* test. It's important to know whether the blood came from a human or an animal such as a pet. The standard test is the *precipitin* test. Injecting human blood into an animal results in the production of antibodies in the animal's bloodstream, and isolating these antibodies from the animal's blood yields an antiserum. If human antiserum creates clotting in a blood sample, the sample must be human.

✔ **If the blood is from an animal, what is the species?** It's possible to create animal antisera in an analogous manner and test for each type of animal.

✔ **If the blood is from a human, what is the blood type?** The procedure for answering this question depends on the sample's quantity and quality. If the quality is good, *direct* typing is done; otherwise, *indirect* typing is used. (We discuss direct typing — to classify blood in the A-B-O system — in more detail in the "Basics of blood typing" sidebar later in the chapter.) A dried bloodstain normally requires indirect typing. The most common indirect typing method is the *absorption-elution* test. Treatment of a sample with antiserum antibodies gives a solution that, upon addition to a known sample, causes coagulation.

✔ **Is it possible to determine the sex, race, and age of the source of the blood?** Here the answers become less precise. Clotting and crystallization indicate age. Testing for testosterone levels and chromosomes can determine sex. And certain controversial, racial genetic markers based on protein and enzyme tests may indicate race.

Other body fluids may contain the same antibodies and antigens found in blood. Therefore, similar tests work on these fluids as well.

In *size exclusion chromatography,* also known as *molecular sieve chromatography, gel filtration chromatography,* or *column chromatography,* a solution passes through a chromatography column filled with porous beads. Molecules that are too large for the pores pass straight through. Molecules that may enter the pores are slowed. The molecules that may enter the pores undergo separation depending on how easily they can enter.

Separating proteins by charge

Methods used to separate proteins by charge include solubility, *ion exchange chromatography,* and *electrophoresis.* Each of these methods is pH dependent.

Proteins are least soluble at their isoelectric point. (The *isoelectric point* is the pH where the net charge on a protein is 0; see Chapter 4 for more info.) At the isoelectric point, many proteins precipitate from solution. At a pH below the isoelectric point, a protein has a net positive charge, whereas a pH above the isoelectric point imparts a net negative charge. The magnitude of the charge depends on the pH and the protein's identity. Therefore, two proteins coincidently having the same isoelectric point don't necessarily have the same net charge at a pH that's one unit lower than the isoelectric point.

Both ion exchange chromatography and electrophoresis take advantage of the net charge. In *ion exchange chromatography,* the greater the magnitude of the charge, the slower a protein moves through a column. This relationship is similar to the ion-exchange process that occurs in water-softening units.

In *electrophoresis,* the sample solution is placed in an electrostatic field. Molecules with no net charge don't move, but species with a net positive charge move toward the negative end and those with a net negative charge move toward the positive end. The net charge's magnitude determines how fast the species move. Other factors influence the rate of movement, but the charge is the key. There are numerous modifications of electrophoresis (including multidimensional movement).

In protein analysis, rarely do biochemists use only one single technique. Two proteins might be similar in size, but differ in charge, and vice versa, because they have a different combination of amino acids. They commonly use several in order to confirm their findings.

Digging into the details: Uncovering a protein's amino acid sequence

When a pure sample of protein is available, you can begin determining its amino acid sequence in order to identify the specific protein. The general procedure for doing so, with slight modification, works for other biochemicals as well.

Step 1: Separating and purifying the polypeptide chains

If you determine that more than one polypeptide chain is present in the protein, you need to separate and purify the chains so you can sequence them individually. (Because many proteins have only one polypeptide chain, this step isn't always necessary.) Denaturing the protein by disrupting its three-dimensional structure without breaking the peptide bonds normally suffices. This can be accomplished by using extremes in pH. If disulfide linkages are present between the chains, apply the procedure outlined in Step 2 to separate the chains for isolation.

Step 2: Slashing intrachain disulfide linkages

Step 2 requires breaking (cleaving) the disulfide linkages. A simple reduction accomplishes this. However, the linkages may reform later, so you need to cleave the linkages and prevent their reformation via reductive cleavage followed by alkylation. Oxidative cleavage, where oxidation of the sulfur to $-SO_3^-$ occurs, also prevents a reversal of the process.

Step 3: Determining amino acid concentration of the chain

Step 3 is easily accomplished using an *amino acid analyzer,* an automated instrument that can determine the amino acid composition of a protein in less than an hour. The instrument requires less than a nanomole of protein. The analyzer's output is the percentages of each of the amino acids present. However, this simply identifies the components present and not their order.

Step 4: Identifying the terminal amino acids

Step 4 not only identifies the terminal amino acids but also indicates whether more than one chain is present. A polypeptide chain only has one N-terminal and one C-terminal amino acid. Therefore, if more than one N- or C-terminal amino acid is present, more than one polypeptide chain must be present.

You can identify the N-terminal residue in a number of ways. In general, procedures begin by adding a reagent that reacts with the N-terminal amino acid and tags it. (Sounds like a schoolyard game.) Subsequent hydrolysis destroys the polypeptide, allowing separation of the tagged residue and its identification. Such methods use Sanger's reagent, dansyl chloride, and leucine aminopeptidase. The method of choice nowadays is called the *Edman degradation.* Like other methods, this method tags the N-terminal residue; however, only the terminal amino acid is cleaved from the chain, so the remainder of the chain is not destroyed as it is in other methods. You can repeat the procedure on the shortened chain to determine the next residue. In principle, repetition of the Edman degradation can yield the entire sequence, but in most cases, determination of the first 30 to 60 residues is the limit.

You can also determine the C-terminal residue by tagging. The akabori reaction (hydrazinolysis) and reduction with lithium aluminum hydride tag the C-terminal residue. You can also selectively cleave the C-terminal residue using the enzyme *carboxypeptidase,* a variety of which are available. Unfortunately, the enzyme doesn't stop with one cleavage; given sufficient time, it proceeds down the entire polypeptide chain. (This reminds me of the Pacman video game.)

Steps 5 and 6: Breaking the chain into smaller pieces

In Step 5, you cleave the polypeptide into smaller fragments and determine the amino acid composition and sequence of each fragment. Step 6 repeats Step 5 using a different cleavage procedure to give a different set of fragments. Steps 5 and 6 break the chain into smaller pieces to ease identification.

Most of the methods here employ enzymes; however, other less-specific methods are useful in some cases. Partial acid hydrolysis randomly cleaves the protein chain into a number of fragments. Trypsin, a digestive enzyme, specifically cleaves on the C-side of arginine or lysine. Using trypsin gives additional information that the total number of arginine and lysine residues present is one less than the number of fragments generated. The digestive enzyme chymotrypsin preferentially cleaves residues containing aromatic rings (tyrosine, phenylalanine, and tryptophan). It slowly cleaves other residues, especially leucine. Clostripain cleaves positively charged amino acids, especially arginine. It cleaves lysine more slowly. Fragments with a C-terminal aspartic acid or glutamic acid form from the interaction of staphylococcal protease on a protein in a phosphate buffer. In the presence of bicarbonate or acetate buffer, only C-terminal glutamic acid fragments result. A number of less specific enzymes can complete the breakdown of the fragments, including elastase, subtilisin, thermolysin, pepsin, and papain.

REAL WORLD

Basics of blood typing

The determination of blood type in the A-B-O system, first begun in 1901, is based on *antigen-antibody reactions.* Over the years, additional reactions have been discovered. More than 256 antigens are known, leading to 23 different blood groups. Each blood group is defined by the antibodies present in the serum and the antigens present on the red blood cells.

In basic blood typing, you need two antiserums, labeled *anti-A* and *anti-B.* Adding a drop of one of these to a blood sample causes coagulation if the appropriate antigens are present. Anti-A interacts with both A and AB blood. Anti-B interacts with both B and AB blood. Neither interacts with type O blood. The approximate distribution of the different blood types is 43 to 45 percent type O; 40 to 42 percent type A; 10 to 12 percent type B; and 3 to 5 percent type AB. Subgrouping is also possible with designations such as O1 and O2, and other very rare types exist as well.

The Rh factor provides an additional means of subdividing blood. The *Rh factor* (the name comes from the rhesus monkey) is an antigen on the surface of red blood cells. A person with a positive Rh factor has a protein (antibody) that's also present in the bloodstream of the rhesus monkey. About 85 percent of humans are Rh positive. A person who lacks this protein is, naturally, Rh negative. Assigning a blood sample as Rh positive or Rh negative is a useful simplification. There are about 30 possible combinations of factors.

Additional factors can determine whether blood belongs to a specific individual: the identification of other proteins and enzymes present in the blood. A forensic serologist (see the "Forensics: Analysis of bloodstains" sidebar earlier in the chapter for more) does this level of testing in every case where the quality of the sample allows. One of the characteristics of proteins or enzymes in the blood is *polymorphism,* or the ability to be present as isoenzymes — that is, the protein may exist in different forms or variants. One well-known example is the polymorphism of hemoglobin into the form that causes sickle cell anemia. Some well-recognized polymorphisms are

Adenyl kinase	AK
Adenosine deaminase	ADA
Erythrocyte acid phosphatase	EAP
Esterase D	EsD
Glucose-6-phosphate Dehydrogenase	G-6-PD
Glutamic pyruvate transaminase	GPT
Phosphoglucomutase	PGM 2-1
6-phosphogluconate Dehydrogenase	6-PGD
Transferrin	Tf

The distribution of each of these *polymorphs* in the population is well established. The determination of each of these additional factors narrows down the number of possible individuals.

Chemical methods of breaking up the fragments include treatment with cyanogen bromide and hydroxylamine, and then heating in an acidic solution. Cyanogen bromide specifically attacks methionine. Hydroxylamine specifically attacks asparagine-glycine bonds. If a solution at a pH of 2.5 is heated to 104 degrees Fahrenheit (40 degrees Celsius), selective cleavage of aspartic acid-proline bonds occurs.

You can apply the Edman degradation technique to each of the fragments. This can simplify the determination of the sequence of a large protein.

Step 7: Combining information to get the total sequence

Step 7 is where the information from the various procedures comes together. It's where you assemble the puzzle using the various parts you've found. For example, look at a simple octapeptide fragment from a protein. This fragment gave, upon complete hydrolysis, one molecule each of alanine (Ala), aspartic acid (Asp), glycine (Gly), lysine (Lys), phenylalanine (Phe), and valine (Val), as well as two molecules of cysteine (Cys). The following fragments were isolated after partial hydrolysis: Gly-Cys, Phe-Val-Gly, Cys-Asp, Cys-Ala, Lys-Cys, and Cys-Asp-Lys. Now you match the fragments, deduce the amino acid sequence in the octapeptide, and write a primary structure for the peptide:

> Cys-Asp Lys-Cys
>
> Cys-Asp-Lys Cys-Ala
>
> Gly-Cys
>
> Phe-Val-Gly
>
> Phe-Val-Gly-Cys-Asp-Lys-Cys-Ala

Step 8: Locating the disulfide linkages

Step 8 doesn't specifically deal with a protein's primary structure, but it is related. If the disulfide linkages are left intact by skipping Step 2, different fragments result. This can be used to determine a protein's overall shape. In some cases, sophisticated instrumental analysis techniques can determine more detailed structural information.

Chapter 6

Enzyme Kinetics: Getting There Faster

*E*nzymes are complex biological molecules, primarily or entirely protein, that behave as biological catalysts. As *catalysts,* they increase the rate of a chemical reaction without themselves being consumed in the reaction. Enzymes are normally very specific in their action, often targeting only one specific reacting species, known as the *substrate.*

This specificity includes *stereospecificity,* the arrangement of the substrate atoms in three-dimensional space. Stereospecificity is illustrated by the fact that if the D-glucose in your diet were replaced by its enantiomer, L-glucose, you wouldn't be able to metabolize this otherwise identical enantiomer. (Makes you want to check what glucose is in that IV fluid bag, doesn't it?)

Enzymes occur in many forms. Some enzymes consist entirely of proteins, whereas others have nonprotein portions known as *cofactors.* The cofactor may be a metal ion, such as magnesium, or an organic substance. An organic cofactor is called a *coenzyme* (there's no specific term for a metallic cofactor). An enzyme lacking its cofactor is an *apoenzyme,* and the combination of an apoenzyme and its cofactor is a *holoenzyme.* A *metalloenzyme* contains an apoenzyme and a metal ion cofactor. A tightly bound coenzyme is a *prosthetic group.* (Wow! We know that this is a lot of terminology, but hang in there. The key is the enzyme.)

One region on the enzyme, the *active site,* is directly responsible for interacting with the reacting molecule(s). When a reacting molecule, the substrate, binds to this active site, a reaction may occur. Other materials besides the enzyme and substrate may be necessary for the reaction to occur.

In many cases, the cell initially produces the enzyme in an inactive form, called a *proenzyme* or *zymogen,* that must undergo activation for it to function. The enzyme trypsin illustrates why generating an inactive form of an enzyme is sometimes necessary. Trypsin is one of the enzymes present in the stomach that's responsible for the digestion of proteins. Its production, as an inactive form, occurs in the cells of the stomach walls, and activation occurs after its release into the stomach. If trypsin were produced in the active form, it would immediately begin digesting the cell that produced it. Eating yourself is not a good thing! Thus, proenzymes and similar structures like prohormones and preprohormones give these proteins a greater spatial and temporal flexibility to their actions.

The activation of the inactive form of an enzyme serves as one form of enzyme control. Inhibition is another method of enzyme control. The two general types of inhibition are *competitive* inhibition, in which another species competes with the substrate to interact with the active site on the enzyme, and *noncompetitive* inhibition, in which the other species binds to some site other than the active site. This binding alters the overall structure of the enzyme so that it no longer functions as a catalyst.

Enzyme Classification: The Best Catalyst for the Job

Ever wonder who gets to name chemicals? Well, the answer varies, but for enzymes it's the Enzyme Commission of the International Union of Biochemistry. Common names for enzymes begin with some description of its action and end with the *-ase* suffix. (Enzymes that were named before the implementation of the -ase system, such as trypsin, don't follow this convention.) The commission has also developed a numerical system for classifying enzymes. The names begin with EC (for Enzyme Commission) and end with four numbers, separated by decimal points, that describe the enzyme. An example of this nomenclature is EC 2.7.4.4.

The first number in the EC name tells you which of the six major enzyme classes the enzyme belongs to (see Table 6-1). For example, the 2 in EC 2.7.4.4 designates the enzyme as a *transferase.* The second number, the 7, indicates what *group* the enzyme transfers. The third number, the first 4, indicates the *destination* of the transferred group. And the last number, the second 4, refines the information given by the third number.

Table 6-1	Six Basic Types of Enzymes
Class of Enzymes	*What They Catalyze*
Oxidoreductases	Redox reactions
Transferases	The transfer groups of atoms
Hydrolases	Hydrolysis
Lyases	Additions to a double bond, or the formation of a double bond
Isomerases	The isomerization of molecules
Ligases or synthetases	The joining of two molecules

We take a more detailed look at each class of enzymes in the following sections.

Up one, down one: Oxidoreductases

Oxidoreductases catalyze a simultaneous oxidation and a reduction. An *oxidation* involves an increase in an element's oxidation state, whereas a *reduction* involves a decrease in the element's oxidation state. (For a more complete discussion of oxidation and reduction, see *Chemistry For Dummies,* written by this book's coauthor, John T. Moore, and published by Wiley.) It's impossible to have one without the other. Table 6-2 lists examples of the types of reactions that qualify as oxidation and reduction. In general, the substrate undergoes either oxidation or reduction, while the enzyme temporarily does the opposite but eventually returns to its original form.

Table 6-2	Some Possible Types of Oxidation and Reduction Reactions
Oxidation	*Reduction*
Loss of one or more electrons	Gain of one or more electrons
Gain of oxygen	Loss of oxygen
Loss of hydrogen	Gain of hydrogen

Here's an example: Succinate dehydrogenase catalyzes the oxidation of the succinate ion. In this case, the oxidation involves the loss of two hydrogen atoms with the formation of a trans double bond. The enzyme alcohol dehydrogenase removes two hydrogen atoms from an alcohol to produce an aldehyde. The general form, unbalanced, of these reactions is as follows:

$$^-OOCCH_2CH_2COO^- \xrightarrow{\text{succinate dehydrogenase}} {}^-OOCCH = CHOO^-$$

$$CH_3CH_2OH \xrightarrow{\text{alcohol dehydrogenase}} CH_3CHO$$

You don't belong here: Transferases

The purpose of a *transferase* is to catalyze the transfer of a group from one molecule to another. *Aminotransferase* transfers an amino group, and *phosphotransferase* transfers a phosphoryl group. The general form, unbalanced, of these reactions appears in Figure 6-1.

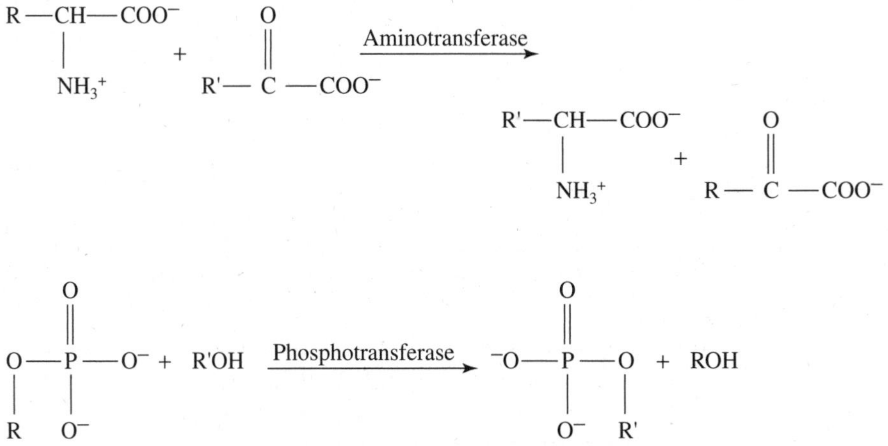

Figure 6-1: General form, unbalanced, of two transferase catalyzed reactions.

Water does it again: Hydrolases

Hydrolases catalyze the cleavage of a bond through the insertion of a water molecule (as an H and an OH). There may be a pH dependence, which results in the subsequent loss of a hydrogen ion. A phosphatase catalyzes the hydrolysis of a monophosphate ester, and a peptidase catalyzes the hydrolysis of a peptide bond. The general form of these reactions appears in Figure 6-2.

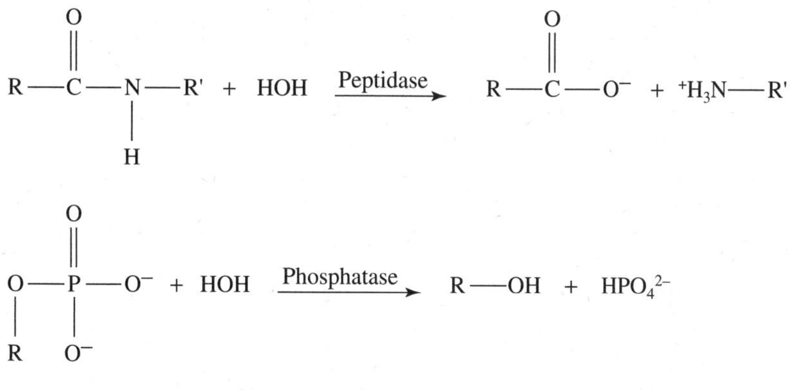

$$R-\overset{\overset{\displaystyle O}{\|}}{C}-\underset{\underset{\displaystyle H}{|}}{N}-R' \ + \ HOH \ \xrightarrow{\text{Peptidase}} \ R-\overset{\overset{\displaystyle O}{\|}}{C}-O^- \ + \ ^+H_3N-R'$$

Figure 6-2:
General form of two hydrolase catalyzed reactions.

$$O-\overset{\overset{\displaystyle O}{\|}}{\underset{\underset{\displaystyle R}{|}}{P}}-O^- \ + \ HOH \ \xrightarrow{\text{Phosphatase}} \ R-OH \ + \ HPO_4^{2-}$$

Taking it apart: Lyases

Lyases catalyze the removal of a group. This process is accompanied by the formation of a double bond or the addition of a group to a double bond. A *deaminase* aids in the removal of ammonia, and a decarboxylase catalyzes the loss of CO_2. The general form of these reactions appears in Figure 6-3.

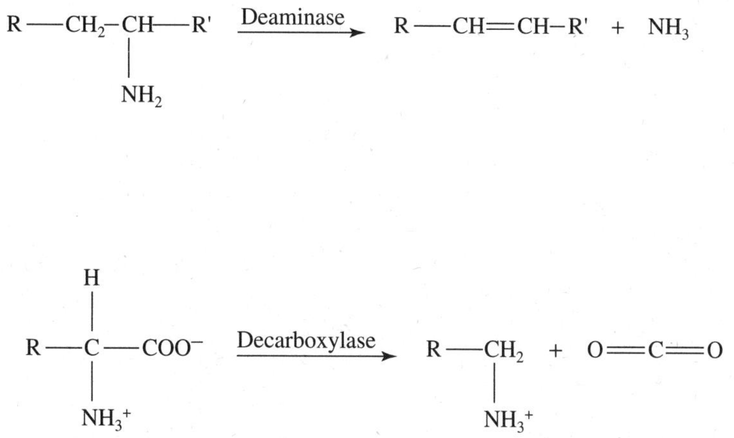

$$R-CH_2-\underset{\underset{\displaystyle NH_2}{|}}{CH}-R' \ \xrightarrow{\text{Deaminase}} \ R-CH{=}CH-R' \ + \ NH_3$$

Figure 6-3:
General form of two lyase catalyzed reactions.

$$R-\overset{\overset{\displaystyle H}{|}}{\underset{\underset{\displaystyle NH_3^+}{|}}{C}}-COO^- \ \xrightarrow{\text{Decarboxylase}} \ R-\underset{\underset{\displaystyle NH_3^+}{|}}{CH_2} \ + \ O{=}C{=}O$$

Shuffling the deck: Isomerases

Racemase and epimerase are isomerases. *Isomerase* enzymes catalyze the conversion of one isomer to another. The *racemase* illustrated at the top of Figure 6-4 catalyzes the racemization of enantiomers. An *epimerase*, like the one at the bottom of Figure 6-4, catalyzes the change of one epimer to another. Like all catalyzed reactions, these are equilibrium processes.

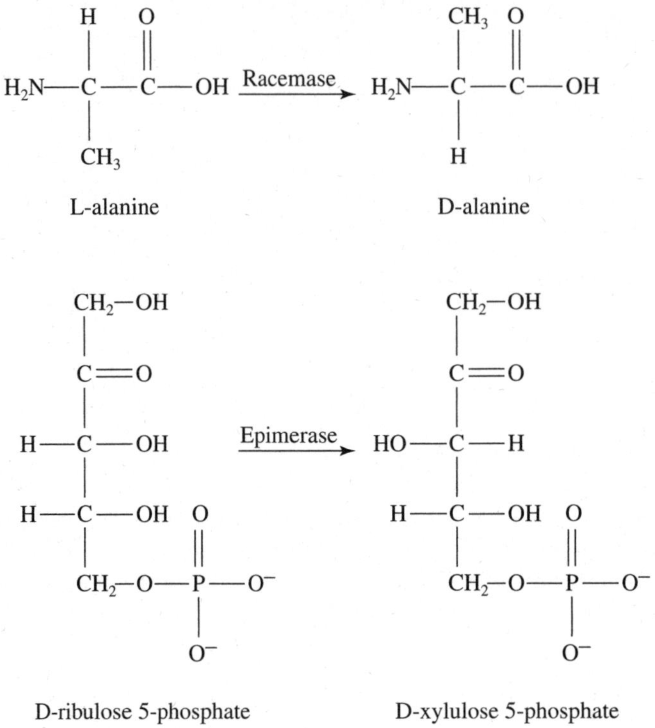
Racemase
L-alanine
D-alanine
Epimerase
D-ribulose 5-phosphate
D-xylulose 5-phosphate

Figure 6-4: Examples of isomerase reactions catalyzed by a racemase and an epimerase.

Putting it together: Ligases

Ligase enzymes catalyze reactions leading to the joining of two molecules in which a covalent bond forms between the molecules. The process often utilizes high-energy bonds such as in ATP. Figure 6-5 illustrates the action of two ligases, pyruvate carboxylase and acetyl-CoA synthetase. *Pyruvate carboxylase* catalyzes the formation of a C-C bond. *Acetyl-CoA synthetase* catalyzes the formation of a C-S bond.

Figure 6-5: Reactions illustrating the action of the ligases pyruvate carboxylase and acetyl-CoA synthetase.

$$\text{Pyruvate} + CO_2 + H_2O + \text{ATP} \xrightleftharpoons{\text{Pyruvate carboxylase}} \text{oxaloacetate} + \text{ADP} + P_i$$

$$\text{Acetate} + \text{CoA-SH} + \text{ATP} \xrightleftharpoons{\text{Acetyl-CoA synthetase}} \text{acetyl-S-CoA} + \text{AMP} + PP_i$$

Enzymes as Catalysts: When Fast Is Not Fast Enough

The action of an enzyme begins with the formation of an *enzyme-substrate complex*. In this formation, the substrate in some way binds to the enzyme's active site. In Chapter 5 we discuss the structure of proteins. The different types of structures (primary, secondary, and so on) are important in the enzyme-substrate interaction. The interaction between the enzyme and the substrate must, in some way, facilitate the reaction, and it opens a new reaction pathway.

The active site is typically a very small part of the overall enzyme structure. The amino acid residues that comprise the active site may come from widely separated regions of the protein (primary structure), and only through interactions leading to higher structure levels are they brought close together. Amino acid residues not in the active site serve many different functions that aid the function of the enzyme.

The first attempt at explaining this process led to the *Lock and Key Model,* in which the substrate behaves as a key that fits into a lock — the enzyme (see Figure 6-6). The Lock and Key Model, to a certain degree, explains the specificity of enzymes. Just as only the right key fits into a lock, only the right substrate fits into the enzyme.

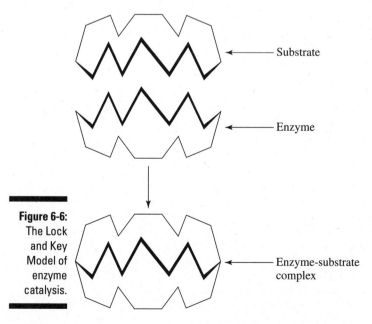

Substrate

Enzyme

Figure 6-6:
The Lock and Key Model of enzyme catalysis.

Enzyme-substrate complex

One limitation of the Lock and Key Model is that it doesn't explain why the reaction actually occurs, and another limitation is that enzymes are flexible and not rigid as this theory implies.

The *Induced-Fit Model* overcomes some of the limitations of the Lock and Key Model. In this model, the substrate still needs to fit into the enzyme like a key, but instead of simply fitting into the "keyhole," some type of modification is induced in the substrate, enzyme, or both. The modification begins the process of the reaction. Figure 6-7 illustrates how the Induced-Fit Model applies to the formation of the same enzyme-substrate in Figure 6-6.

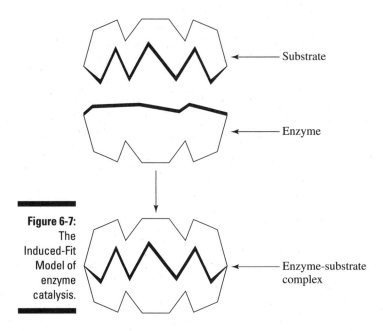

Substrate

Enzyme

Figure 6-7:
The Induced-Fit Model of enzyme catalysis.

Enzyme-substrate complex

All about Kinetics

As you know, all reactions involve energy. The reactants begin with a certain level of energy, an additional quantity of energy is absorbed to reach the transition state ($\triangle G^*$, where the asterisk indicates the transition state), and then energy is released to reach the products. The difference in the energy between the reactants and products is $\triangle G$.

If the energy level of the products is greater than that of the reactants (energy is absorbed), the reaction is *endergonic* and *nonspontaneous*. If the energy level of the products is less than that of the reactants (energy is released), the reaction is *exergonic* and *spontaneous*.

But just because a reaction is spontaneous doesn't mean it occurs at an appreciable rate. The rate depends on the value of $\triangle G^*$. The greater the value of $\triangle G^*$, the slower the reaction is. An enzyme, like any catalyst, lowers the value of $\triangle G^*$ and consequently increases the reaction's rate. The difference between the reactants and products remains unchanged, as does their equilibrium distribution. The enzyme facilitates the formation of the transition state (see Figure 6-8).

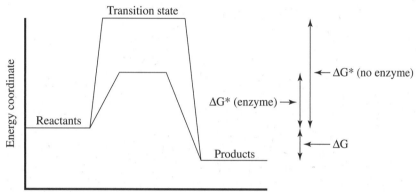

Figure 6-8:
Effect of an enzyme on a reaction.

A species has two possible fates in the transition state (if only life were that simple for humans): It may lose energy and return to the reactant form, or it may lose energy and move to the product form. These two fates lead to two equilibria. One of the equilibria involves the reactant (substrate) and the transition state, and the other involves the product(s) and the transition state. Rapid removal of the product(s) doesn't allow establishment of the reverse process that leads to the equilibrium. Removal of the product(s) simplifies the analysis of the kinetic data.

Enzymes, like all catalysts, catalyze both the forward and the reverse reactions. The lowering of $\triangle G^*$ accelerates both reactions. The ultimate equilibrium concentrations of substrate and products are the same whether an enzyme is present or not; the enzyme merely changes the amount of time necessary to reach this state.

Enzyme assays: Fixed time and kinetics

An *enzyme assay* is an experiment to determine an enzyme's catalytic activity. One can measure either the rate of disappearance of the substrate or the rate of appearance of a product. The experimental mode of detection depends on the particular chemical and physical properties of the substrate or the product, and the rate is the change in concentration per change in time. In a *fixed-time assay,* you simply measure the amount of reaction in a fixed amount of time. In a *kinetic assay,* you monitor the progress of a reaction continuously. After you determine the rate of change in concentration of any reactant or product, you can determine the rate of change for any other reactant or product of the reaction.

 Controlling the conditions precisely is important. Minor changes in variables such as the temperature or the pH can drastically alter an enzyme's catalytic activity. For example, you should carry out the study of enzymes important to humans at 98.6 degrees Fahrenheit (37 degrees Celsius) because this is normal body temperature.

Rate determination: How fast is fast?

Controlling kinetics experiments closely is critical. After you determine the basic conditions, you can run a series of experiments using a fixed enzyme concentration and varying concentrations of substrate. Up to a point, an increase in substrate concentration results in an increase in reaction rate. The rate increases until the enzyme is saturated. This *saturation point* is where all the enzyme molecules are part of an enzyme-substrate complex. When this occurs, an increase in the substrate concentration yields no increase in the rate because no enzymes are available to interact with the additional substrate molecules. For most reactions, the rate of the reaction approaches the saturation level along a hyperbolic curve. Theoretically, the reaction rate only reaches saturation at infinite substrate concentration.

A plot of the reaction rate, V, versus the substrate concentration, [substrate], supplies several bits of useful (and probably some not so useful) data (see Figure 6-9). The experiment is at constant enzyme concentration. One piece of useful data is the maximum reaction rate, V_{max}. The rate approaches V_{max} asymptotically. At low substrate concentrations, the reaction approaches first-order kinetics, where the rate of reaction depends only on the concentration of one reactant. At high concentrations, the reaction approaches zero-order kinetics, where the rate of reaction is independent of reactant concentration. (Later in this chapter you'll see that this graph varies with less simple enzyme-substrate interactions.) In the region between the zero-order region and the first-order region, the kinetics are mixed and difficult to interpret. Important values in the low-concentration region (first-order region) are ½ V_{max} and K_M. K_M is the *Michaelis constant,* which corresponds to the substrate concentration producing a rate of ½ V_{max}. The Michaelis constant, measured in terms of molarity, is a rough measure of the enzyme-substrate affinity. K_M values vary widely. (Okay, take a deep breath and try not to run screaming from the room!)

At low substrate concentrations, the relationship between [substrate] and V is approximately linear. At high substrate concentrations, though, V is nearly independent of [substrate]. The low substrate region is useful in the application of the Michaelis-Menten equation (see the next section).

In an uncatalyzed reaction, increasing the substrate concentration doesn't lead to a limiting V_{max}. The rate continues to increase with increasing substrate concentration. This indirect evidence leads to the conclusion that there's an *enzyme-substrate complex,* a tightly bound grouping of the enzyme and the substrate. The limit occurs when all the enzyme molecules are part of a complex and no free enzyme molecules are available to accommodate the additional substrate molecules. Various X-ray and spectroscopic techniques provide direct evidence to confirm the formation of an enzyme-substrate complex.

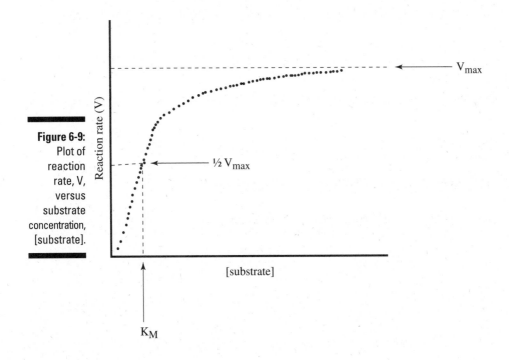

Figure 6-9: Plot of reaction rate, V, versus substrate concentration, [substrate].

Enzymes in medical diagnosis and treatment

Because enzyme levels can indicate medical problems, enzyme assays are useful for both the diagnosis and treatment of medical problems. For example, creatine kinase (CK) is an enzyme that aids in the synthesis and degradation of creatine phosphate.

CK exists as three different isoenzymes. Each is composed of two polypeptide chains. In the case of muscle CK, labeled CK-MM, the chains are identical. The CK found in the brain, labeled CK-BB, also has identical polypeptide chains, but they're different from the ones associated with muscle CK. Finally, the CK found in the heart is a hybrid of the two, with one M chain and one B chain: CK-MB.

Normal blood serum contains a little CK-MM and almost no CK-BB or CK-MB. When a tissue is injured, though, some of the intracellular enzymes leak into the blood, where they can be measured. Elevated levels of total CK (all three isoenzymes) may be indicative of skeletal-muscle trauma or myocardial infarction (MI, or heart attack). Analysis of the individual isoenzymes may give additional clues.

For example, say an individual falls off a ladder and suffers several broken bones. He is taken to the hospital, where his blood serum CK is measured. It's elevated as expected, but the physician also orders a CK-MB level determination. The CK-MB level turns out to also be highly elevated, indicating that the reason the man fell off the ladder to begin with was that he was suffering a heart attack (CK-MB). This knowledge allows the doctor to start a treatment regimen that helps to minimize permanent heart damage.

Measuring Enzyme Behavior: The Michaelis-Menten Equation

One of the breakthroughs in the study of enzyme kinetics was the development of the Michaelis-Menten equation. Biochemists can apply the equation to kinetic data to interpret the behavior of many enzymes. (There are exceptions, however, and they don't give a graph like the one in Figure 6-9.) In general, the results of the kinetics experiments are for *allosteric* (regulatory) enzymes. The *Michaelis-Menten equation* is as follows:

$$V = \frac{V_{max}[S]}{[S]+K_M}$$

In this equation, V is the rate of the reaction, [S] is the substrate concentration, V_{max} is the maximum reaction rate, and K_M is the Michaelis constant. As seen in Figure 6-9, the rate of catalysis, V, increases linearly at low substrate concentration but begins to level off at higher concentrations. Interpretation begins with examining the following general reaction pathway:

$$E + S \underset{k_{-1}}{\overset{k_1}{\rightleftharpoons}} ES \underset{k_{-2}}{\overset{k_2}{\rightleftharpoons}} E + P$$

In this pathway, E is the enzyme, S is the substrate, ES is the enzyme-substrate complex, and P is the product. The various instances of k refer to the rate constants of the various steps; a negative rate constant is for the reverse process. In the first step, the separate enzyme and substrate combine to form the enzyme-substrate complex (transition state). The rate of formation of ES is k_1. After ES forms, it may break down to E and $S(k_{-1})$ or it may proceed to product (k_2). (**Note:** Some texts refer to k_2 as k_{cat}.)

Because the enzyme will catalyze the reverse process, E and P may combine to reform the complex (k_{-2}). Ignoring the reverse reaction (k_{-2}) simplifies the interpretation of the data. This isn't an unreasonable assumption if data collection is near the beginning of the reaction, where the concentration of P is low. The assumption that k_{-2} is negligible leads to a simplification of the preceding equation to:

$$E + S \underset{k_{-1}}{\overset{k_1}{\rightleftharpoons}} ES \overset{k_2}{\longrightarrow} E + P$$

Through this simplification, the chemists Leonor Michaelis and Maud Menten were able to propose a model that explains the kinetics of many different enzymes. Through their work, an expression relating the catalytic rate to the concentrations of the enzyme and substrate and to the individual rates was developed. The starting point for this expression is the relationship between the rate of the reaction and the concentration of the enzyme-substrate complex:

$$V = k_2[ES]$$

Similarly, the rate of formation of ES is $k_1[E][S]$, and the rate for the breakdown of ES is $(k_{-1} + k_2)[ES]$. Throughout most of the reaction, the concentration of ES remains nearly constant. This is the *steady-state assumption,* which assumes that during a reaction the concentrations of any intermediates remain nearly constant. This assumption means that the rate of formation of ES must be equal to the rate of breakdown of ES, or:

$$k_1[E][S] = (k_{-1} + k_2)[ES]$$

This equation rearranges to:

$$\frac{[E][S]}{[ES]} = \frac{(k_{-1} + k_2)}{k_1} = K_M$$

The combination of the three rate constants yields a new constant: the Michaelis constant, K_M, which is independent of the actual concentration of the enzyme and substrate; however, it is related to the ratio of the concentrations. For this reason, K_M is an important characteristic of enzyme-substrate interactions. Using the Michaelis constant, the concentration of ES is:

$$[ES] = \frac{[E][S]}{K_M}$$

When the enzyme concentration is much lower than the substrate concentration, the value of [S] is very close to the total substrate concentration. The enzyme concentration, [E], is equal to the total enzyme concentration, $[E]_T$, minus the concentration of the enzyme-substrate complex, or $[E] = [E]_T - [ES]$. If you enter this relationship into the preceding equation, you get:

$$[ES] = \frac{([E]_T - [ES])[S]}{K_M}$$

Rearranging this equation gives

$$[ES] = \frac{[E]_T / K_M}{1 + [S]/K_M} = \frac{[E]_T [S]}{[S] + K_M}$$

Substituting this relationship into $V = k_2[ES]$ or ($V = k_{cat}[ES]$) gives

$$V = k_2 [E]_T \frac{[S]}{[S] + K_M}$$

The maximum rate, V_{max}, occurs when all the enzyme molecules are associated with substrate. That is, $[ES] = [E]T$. This changes $V = k_2[ES]$ to $V_{max} = k_2[E]_T$. This relationship changes the preceding equation to the Michaelis-Menten equation:

$$V = \frac{V_{max}[S]}{[S] + K_M}$$

This equation accounts for the information depicted in Figure 6-9. At very low concentrations, $[S] \ll K_M$, $V = (V_{max}/K_M)[S]$, and when [S] is greater than K_M (high [S]), $V = V_{max}$. When $[S] = K_M$ it leads to $V = V_{max}/2$.

Ideal applications

The Michaelis-Menten equation explains the behavior of many enzymes. Determining both the K_M and V_{max} values is relatively easy and is normally done graphically using computer programs that generate the best-fit curve.

The K_M values vary widely. The value depends on the identities of the enzyme and substrate and on a variety of environmental factors such as temperature, ionic strength, and pH. Because K_M indicates the substrate concentration required to fill half of the active sites on the enzyme, it also indicates the minimum substrate concentration for significant catalytic activity to occur. You can determine the fraction of sites filled, f_{ES}, from the value of K_M:

$$f_{ES} = \frac{V}{V_{max}} = \frac{[S]}{[S] + K_M}$$

K_M also gives information about the rate constants for the reaction:

$$\frac{(k_{-1} + k_2)}{k_1} = K_M$$

When k_{-1} is significantly greater than k_2, $K_M = k_{-1}/k_1$, which relates to the equilibrium constant for the dissociation of the enzyme-substrate complex:

$$K_{ES} = \frac{[E][S]}{[ES]} = \frac{k_{-1}}{k_1}$$

K_M is a measure of the binding in the enzyme-substrate complex. A high K_M value indicates that the binding is weak, whereas a low value indicates that the binding is strong.

The value of V_{max} supplies the enzyme's *turnover number* — the number of substrate molecules transforming to products per unit of time for a fully saturated enzyme. You can determine k_2 from this value. (The constant k_2 is also known as the *catalytic constant,* k_{cat}.) If the concentration of active sites, $[E]_T$, is known, this relationship applies:

$$V_{max} = k_2[E]_T$$

And:

$$k_2 = V_{max}/[E]_T$$

Realistic applications

The ideas in the preceding section provide useful information about the behavior of many enzymes. In cells, however, the enzymes are seldom saturated with substrate. Under typical conditions, $[S]/K_M$ is usually between 1.0 and 0.01. If K_M is much greater than $[S]$, the catalytic rate, k_{cat} (or k_2), is significantly less than the ideal value because only a small portion of the active sites contains substrate. The ratio k_{cat}/K_M allows you to compare the substrate preferences of an enzyme.

The maximum rate of catalytic activity is limited by the rate of diffusion to bring the enzyme and substrate together. Some enzymes can exceed this limit by forming *assemblages.* In these groups, the product of one enzyme is the substrate for a closely associated enzyme. This allows a substrate to enter the group and pass from enzyme to enzyme as if it were in an assembly line. (Henry Ford would be proud.)

Another complication is that many enzymes require more than one substrate. It's possible to utilize these multiple substrates through sequential displacement or through double displacement. In *sequential displacement,* all substrates must simultaneously bind to the enzyme before the release of the product. In this type of displacement, the order in which the substrates bind is unimportant. In *double displacement,* or *ping-pong* situations, one or more products leave before all the substrates bind. Double displacement mechanisms temporarily modify the enzyme.

Here we go again: Lineweaver-Burk plots

Once upon a time, before the invention of computers, the determination of K_M and V_{max} was a tedious process. Today, curve-fitting programs allow rapid analysis of the data to determine these values. However, a relatively simple method (if there can be such a thing in biochemistry) allows a ballpark determination of these two constants. This method is to construct a *Lineweaver-Burk plot,* also known as a *double-reciprocal plot.* The basis of a Lineweaver-Burk plot comes from the manipulation of the Michaelis-Menten equation to the form

$$\frac{1}{V} = \frac{K_M}{V_{max}} \times \frac{1}{[S]} + \frac{1}{V_{max}}$$

There are problems with a Lineweaver-Burk plot. One problem is that the procedure artificially overemphasizes low substrate velocities. For this reason, small errors in the measurements become large errors in the plot. Because of this, the plot only gives a general idea of K_M and V_{max}. Today, computers can fit the data to a hyperbola, which gives better results.

This equation has the form $y = mx + b$ and describes a straight line. A plot of the reciprocal of the rate, $1/V$, versus the reciprocal of the substrate concentration, $1/[S]$, gives a line with a *y*-intercept equal to $1/V_{max}$ and an *x*-intercept of $-1/KM$. An example of this type of plot appears in Figure 6-10.

The Lineweaver-Burk plot is the most widely used graphical technique for the determination of K_M and V_{max}. However, there are other methods. The *Woolf plot,* shown in Figure 6-11, uses the equation

$$\frac{[S]}{V} = \frac{1}{V_{max}} \times [S] + \frac{K_M}{V_{max}}$$

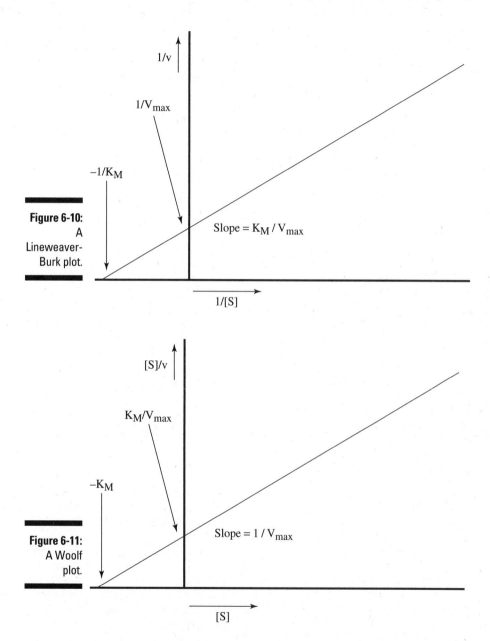

Figure 6-10: A Lineweaver-Burk plot.

Figure 6-11: A Woolf plot.

Plotting [S]/V versus [S] gives a straight line. An *Eadie-Hofstee plot,* shown in Figure 6-12, uses the equation

$$V = -K_M \times \frac{V}{[S]} + V_{max}$$

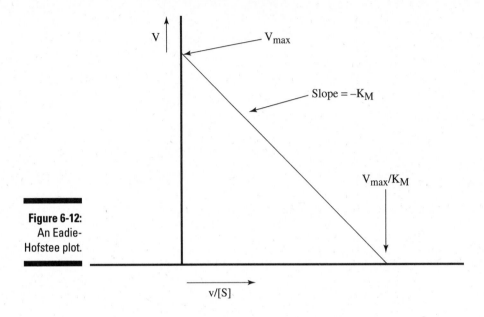

Figure 6-12:
An Eadie-
Hofstee plot.

Plotting V versus V/[S] gives a straight line. (And you know how chemists and mathematicians love a straight line!)

Enzyme Inhibition: Slowing It Down

Inhibitors are substances that decrease an enzyme's activity, and they come in two general classes: *competitive* inhibitors, which compete with the substrate, and *noncompetitive* inhibitors, which don't compete. (Sounds like they're participants in a reality show that takes place on a desert island.) *Mixed* inhibition has characteristics of both competitive and noncompetitive inhibition. (Uncompetitive inhibitors are even more distinct and rare.) In general, these processes are reversible, but there are also irreversible inhibitors that permanently alter the enzyme or bind very strongly to the enzyme. All inhibition may serve as a method of regulating enzymatic activity. This form of inhibition also has many medical applications. Examples include antiepileptic and chemotherapy drugs, along with the ever-popular Viagra. The action of many poisons also occurs through inhibition.

Competitive inhibition

A competitive inhibitor enters the active site of an enzyme and, thus, prevents the substrate from entering. This prevention results in a decrease in the number of enzyme-substrate complexes that form and, hence, a decrease in the rate of catalysis. In most cases, a portion of the inhibitor mimics a portion of the substrate. An increase in the substrate concentration overcomes this inhibition because of the increased probability that a substrate molecule, rather than an inhibitor molecule, will enter the active site.

Noncompetitive inhibition

Noncompetitive inhibitors don't enter the active site but instead bind to some other region of the enzyme. These species usually don't mimic the substrate. This type of inhibitor reduces the enzyme's turnover number. Unlike competitive inhibition, an increase in the substrate doesn't overcome noncompetitive inhibition. This type of inhibition takes many different forms, so there's no simple model.

Graphing inhibition

Lineweaver-Burk plots are useful in the study of enzyme inhibition. Figures 6-13 and 6-14 illustrate how the graph changes in the presence of a noncompetitive and a competitive inhibitor. The plot of enzyme inhibition allows you to quickly determine the type of inhibition. In noncompetitive enzyme inhibition, the value of K_M remains unchanged. In competitive inhibition, however, it's V_{max} that remains unchanged.

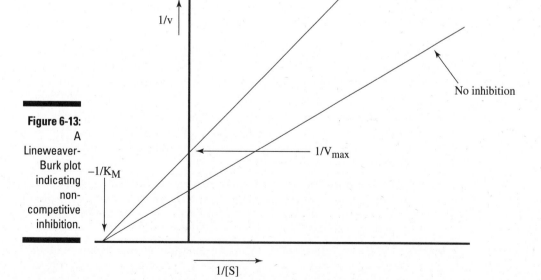

Figure 6-13:
A Lineweaver-Burk plot indicating non-competitive inhibition.

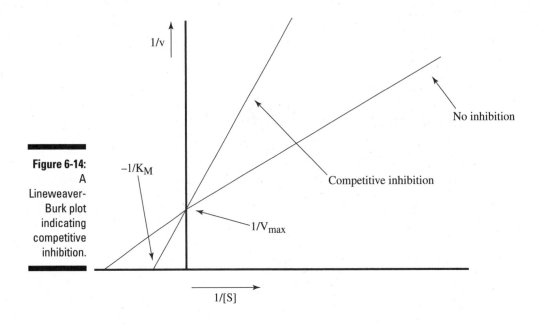

Figure 6-14:
A
Lineweaver-
Burk plot
indicating
competitive
inhibition.

Enzyme Regulation

In general, an increase in the concentration of a substrate, if unregulated, induces an increase in the rate of reaction. An increase in the concentration of a product, in general, has the reverse effect. Product regulation is a type of feedback control. In many cases, regulating the activity of enzymes more precisely is necessary. Enzyme regulation happens in four ways:

✔ **Allosteric control:** An allosterically regulated enzyme has a regulatory site. When a small molecule, called a *regulator,* binds to the regulatory site, it induces a conformational change in the enzyme, making it into its active form.

✔ **Multiple enzyme forms:** Some enzymes have multiple forms known as *isozymes* or *isoenzymes.* These forms have slight differences in their structures. These differences lead to differences in the K_M and V_{max} values and, therefore, in the general activity.

✔ **Covalent modification:** In this form of regulation, the attachment of a group, often a phosphoryl group, alters the enzyme's activity. This process is a reversible form of control. Protein kinases catalyze this type of activation, whereas other enzymes catalyze deactivation.

✔ **Proteolytic activation:** In this form of regulation, an inactive form of an enzyme — a proenzyme or a zymogen — undergoes irreversible conversion to the active form, often through the hydrolysis of one or more peptide bonds.

Where the money is: Enzymes and industry

The industrial implementation of enzymes originated from studies in the food, wine, and beer industries. (These are three of our very favorite industries in which we've invested a great deal of time and money.) Scientists, such as Louis Pasteur, laid much of the groundwork for these applications.

Many of the applications of enzymes to industry involve *immobilized enzymes,* which covalently bond to an insoluble matrix such as cellulose or glass beads. The immobilization of an enzyme stabilizes it and allows prolonged use. Some useful commercial enzymes are as follows:

Carbohydrases

Amylase: Used as a digestive aid for precooked food

Amyloglucosidase: Converts starch to dextrose

Cellulase and hemicellulase: Used in the conversion of sawdust to sugar and the production of liquid coffee concentrates

Glucose isomerase: Used in the production of fructose from cornstarch

Glucose oxidase: Removes glucose from egg solids

Invertase: Stabilizes sugars in soft-centered candy

Lactase: Prevents the crystallization of lactose in ice cream

Pectinase: Clarifies wine and fruit juice

Catalase

Removes H_2O_2 used in the "cold pasteurization" of milk

Proteases

Alcalase: Added to detergents for removal of protein stains

Bromelain: Tenderizes meat

Ficin, streptodornase, and trypsin: Debride wounds

Lipase: Produces flavor in cheese

Lipoxygenase: Whitens bread

Papain: Tenderizes meat and stabilizes beer

Pepsin: Used as a digestive aid for precooked food

Rennin: Used in cheese making

Part III

Carbohydrates, Lipids, Nucleic Acids, and More

The 5th Wave   By Rich Tennant

In this part . . .

We go over many biochemical species. Beginning with carbohydrates, we move on to perhaps less tasty-sounding fare: lipids and steroids. Next up are nucleic acids and that amazing encyclopedia about you that sits on the shelf inside every one of your cells — the genetic code of life, guest starring DNA and RNA. After that we end by talking about vitamins and hormones.

Chapter 7

What We Crave: Carbohydrates

*A*dmit it: You love your carbohydrates. From simple sugars to complex carbohydrates, a day without carbs is a boring day. And carbs are plentiful. In terms of mass, carbohydrates are the most abundant biochemical.

Carbohydrates are a product of *photosynthesis,* where inorganic carbon dioxide becomes organic carbon with the utilization of solar energy, accompanied by the release of oxygen gas. The conversion of solar energy to chemical energy produces carbohydrates, which are the primary energy source for metabolic processes. Carbs not only are an important energy source but also are the raw materials for the synthesis of other biochemicals. They have structural uses and are a component of nucleic acids.

The term *carbohydrate* originally referred to "hydrates of carbon" because the general formula of these compounds was $C_nH_{2n}O_n$ or $C_n(H_2O)_n$. However, some materials with this general formula are not carbohydrates, and some carbohydrates don't have this general formula. Defining carbohydrates as *polyhydroxyaldehydes* and *polyhydroxyketones and their derivatives* is better (though not much more conversational).

Natural carbohydrates are subdivided into *monosaccharides,* or simple sugars containing three to nine carbon atoms; *polysaccharides,* or polymers of monosaccharides; and an intermediate category of *oligosaccharides,* with two to ten monosaccharide units joined. The most important oligosaccharides to humans economically and biologically are the *disaccharides.* Read on to find out more about all these categories of carbohydrates.

Properties of Carbohydrates

In this section, we show you some of the general physical properties of carbohydrates as well as some of their chemical properties.

In general, the names of most carbohydrates are recognizable by an *-ose* suffix. An *aldose,* for example, is a monosaccharide in which the carbonyl group is an aldehyde, whereas in a *ketose,* the carbonyl group is a ketone. Chemists also use roots that refer to the number of carbon atoms. *Pentoses,* with five carbon atoms, and *hexoses,* with six carbon atoms, are very important because they're commonly found in nature and are involved in many important biochemical reactions. *Trioses, tetroses,* and so on are also found in nature. You can combine these generic names to get terms such as *aldohexose* and *ketopentose.*

They contain one or more chiral carbons

Chiral carbons are those that have four different groups, atoms, or groups of atoms attached to them. Most carbohydrates contain one or more chiral carbons. For this reason, they're *optically active:* They rotate polarized light in different directions and often have different activity in biological systems. *Fischer projections* are useful in indicating the asymmetry around each of the chiral carbon atoms. (Figure 7-1 illustrates the construction of a Fischer projection; see Chapter 3 for more info.) In the Fischer projection, the vertical lines project back, and the horizontal lines project forward. Two groups are arranged around a chiral center. These arrangements are called *enantiomers,* and they represent nonsuperimposable mirror images, like left and right gloves. The enantiomers comprise a D/L pair, where the D- form rotates polarized light to the right and the L- form rotates polarized light to the left.

Figure 7-1:
The relationship between the three-dimensional structure around a chiral center and the Fischer projection.

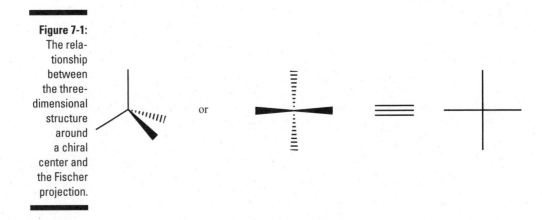

or

TIP

Fischer projections are useful not only for representing chiral carbons but also for identifying which enantiomeric form is present in a sample. To determine whether two projections are enantiomers or simply two representations of the same molecule, you must compare the two drawings. During this comparison, rotate one of the projections by 180 degrees on an axis perpendicular to the plane of the paper (in other words, turn the paper while it's lying on a table). If the diagrams are identical after this rotation, then they're simply two representations of the same molecule. If the diagrams aren't identical, they represent a pair of enantiomers.

They have multiple chiral centers

Because many carbohydrates have more than one chiral center (more than one chiral carbon), there can be more than two *stereoisomers.* The number of stereoisomers is 2^n, where n is the number of chiral carbons. For example, if a compound has two chiral carbons, it has a total of four stereoisomers — two pairs of enantiomers. Although the members of each pair are enantiomers, members of the different pairs are referred to as *diastereomers.*

The structure for *D-glucose,* a typical monosaccharide, appears in Figure 7-2. In this figure (a Fischer projection), all the carbon atoms except the ones at the top and bottom are chiral — a common way of representing monosaccharides. The carbon atoms appear as a vertical chain with the carbonyl carbon as near the top as possible (it's at the top for an aldose). Numbering the carbon atoms begins at the top, as indicated with the top carbon labeled C_1. The highest-numbered chiral carbon in this case is number 5. By convention, if the –OH on this carbon atom appears on the right, it's the D- form of the monosaccharide; if it's on the left, it's the L- form.

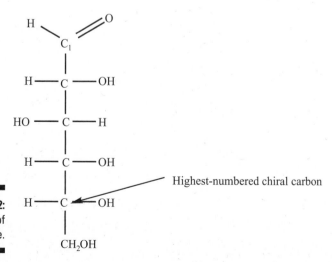

Highest-numbered chiral carbon

Figure 7-2:
Structure of
D-glucose.

Any change in the relative positions of the groups attached to any of the chiral carbon atoms in a Fischer projection produces either a different enantiomer or a diastereomer (assuming that the result isn't simply a different way of drawing the original structure). D-glucose, with 4 chiral centers, has 16 structures. One is D-glucose, and another is its enantiomer, *L-glucose*. The remaining 14 structures are diastereomers consisting of 7 enantiomeric pairs. Each of the enantiomeric pairs consists of a different monosaccharide. In the case of glucose, you have *allose, altrose, glucose, mannose, gulose, idose, galactose,* and *talose,* as shown in Figure 7-3. The different D-ketohexoses are in Figure 7-4.

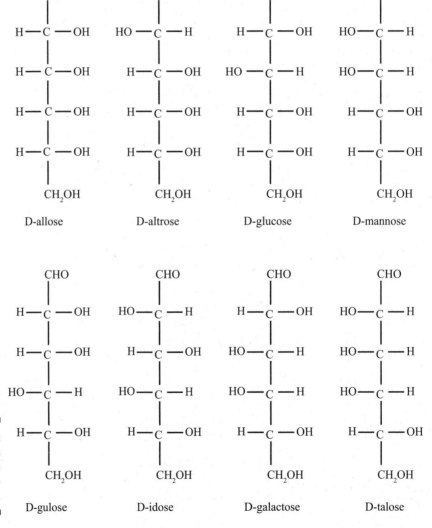

Figure 7-3:
Structures
of the
D-aldo-
hexoses.

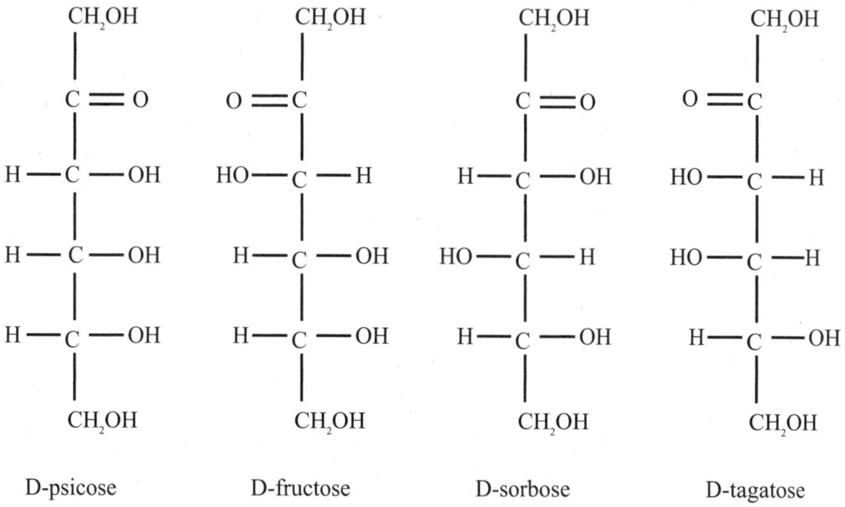

Figure 7-4: Structures of the D-keto-hexoses.

D-psicose   D-fructose   D-sorbose   D-tagatose

A Sweet Topic: Monosaccharides

The *monosaccharides,* or simple sugars, are an important class of biochemicals. For example, glucose, one of the most common monosaccharides, is the primary form of energy storage in the body. Most monosaccharides taste sweet. The relatively large number of hydroxyl groups and the polar carbonyl group mean that most of these compounds are water-soluble. And, as mentioned earlier, most are optically active. (And you'd be active too after eating a bunch of monosaccharides.) The following sections discuss the structures, properties, and derivatives of monosaccharides.

The most stable monosaccharide structures: Pyranose and furanose forms

The most important monosaccharide is D-glucose (one form of D-glucose appears back in Figure 7-2). D-glucose is important because it's the building block of many other carbohydrates. This form exists in equilibrium with two slightly different ring forms. The ring form results from an internal cyclization reaction, where two groups on the same molecule join, forming a ring. (The rings appear as planar structures even though the actual structures aren't planar.) This cyclization involves a reaction between the carbonyl group and the highest-numbered chiral carbon, producing one of the following structures: a *hemiacetal,* an *acetal,* a *hemiketal,* or a *ketal.* In the case of D-glucose, a pyranose ring forms (see Figure 7-5). Haworth projection formulas are especially useful when representing the ring forms of a monosaccharide because Fischer projections would be cumbersome because of the ring system.

Figure 7-5:
A pyranose
ring.

The pyranose structure of D-glucose (see Figure 7-6) and other monosaccharides has two possible structures. If you examine the Fischer projection for D-glucose, you can see why.

- ✔ **Structure 1:** Hydroxyl group on one carbon (the rightmost carbon) in the up position.

- ✔ **Structure 2:** Hydroxyl group on the corresponding carbon in the down position.

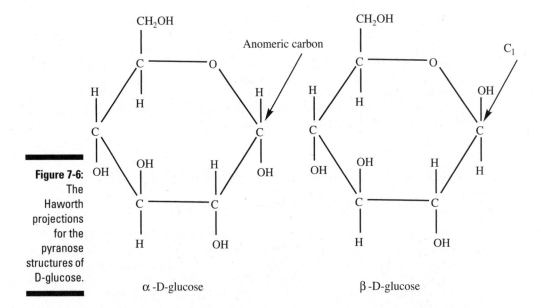

Figure 7-6:
The
Haworth
projections
for the
pyranose
structures of
D-glucose.

If you "bend" the carbonyl group around and then allow a reaction with the highest numbered chiral carbon, you have two choices: right or left. This gives you two forms known as anomers. The anomers are labeled α and β.

The carbonyl carbon — C_1, in this case — is the anomeric carbon, which should be on the right side of a Haworth projection. The relative positions of –H and –OH about the anomeric carbon determine whether it's the α or β form. The hydroxyl group points down in the α form, and the hydroxyl group points up in the β form. (Reversing the drawing of the rings may show a structure with the opposite orientation of the groups about the anomeric carbon.) In solution, each of the anomers is in equilibrium with the open chain form represented by the Fischer projection. Therefore, there's an inter-conversion between the α and β forms known as mutarotation.

It's also possible to form a five-membered ring, called a *furanose* ring. A simplified furanose structure appears in Figure 7-7. Ribose is an example of a monosaccharide that may form a furanose ring.

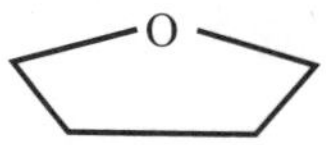

Figure 7-7:
A furanose
ring.

The pyranose and furanose forms are the thermodynamically more stable forms of the monosaccharides. In general, in the equilibria involving ring and open forms, less than 10 percent of the molecules are in the open form. *Fructose* is a ketose that may form a furanose ring. Structures of D-fructose are shown in Figure 7-8.

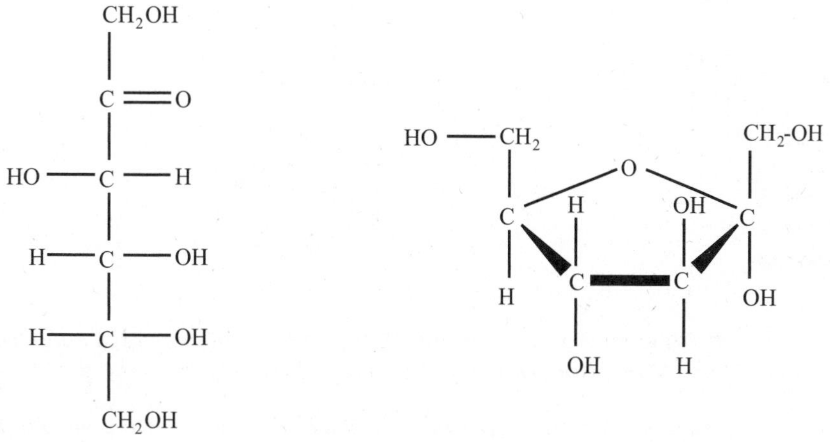

CH$_2$OH

C═O

HO─C─H

H─C─OH

H─C─OH

CH$_2$OH

D-fructose

HO─CH$_2$ CH$_2$-OH

α-D-fructose

Figure 7-8:
Two
forms of
D-fructose.

Chemical properties of monosaccharides

The reaction of a monosaccharide with methanol, CH_3OH, in the presence of hydrochloric acid, HCl, replaces the hydrogen atom of the hydroxyl group on C_1 with a methyl group, forming a glycosidic bond. (Nitrogen may also be part of a glycosidic bond.) After the glycoside forms, the ring is "locked," meaning it won't reopen; therefore, mutarotation no longer takes place. A formerly reducing sugar is no longer a reducing sugar.

Many aldoses, because of the aldehyde group, are reducing sugars — that is, they're reducing agents in certain redox reactions. A number of tests for reducing sugars include using Fehling's solution or Benedict's solution. These tests are useful to check for glucose in the urine of a diabetic (see Chapter 3).

Derivatives of monosaccharides

A variety of derivatives of the monosaccharides are formed through the alteration of one or more of the functional groups present. In this section we examine some of these derivatives using D-ribose as the parent monosaccharide. Two forms of the structure of D-ribose appear in Figure 7-9.

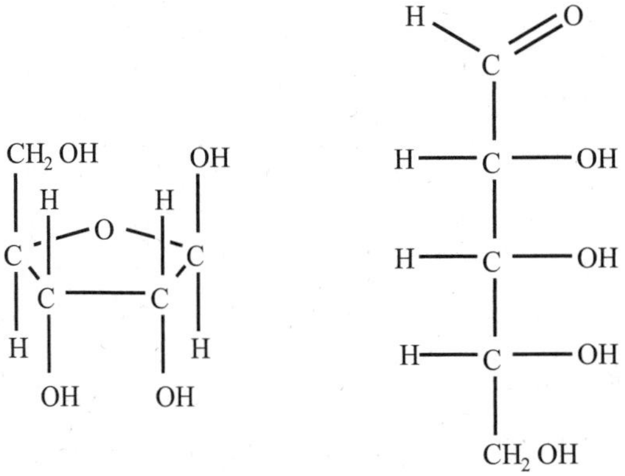

Figure 7-9: Two representations of the structure of D-ribose.

The reduction of the carbonyl group to an alcohol yields a reduced sugar *(polyhydric alcohol)*. The reduction of D-ribose forms *ribitol* (see Figure 7-10).

Oxidizing a monosaccharide to a carboxylic acid is also possible. Two such important oxidations are oxidation of an aldehyde (aldose) to an *aldonic acid* and oxidation of the alcohol on the highest-numbered carbon atom to a *uronic acid*. In the case of D-ribose, it's possible to form D-ribonic acid (Figure 7-11) or D-ribouronic acid (Figure 7-12).

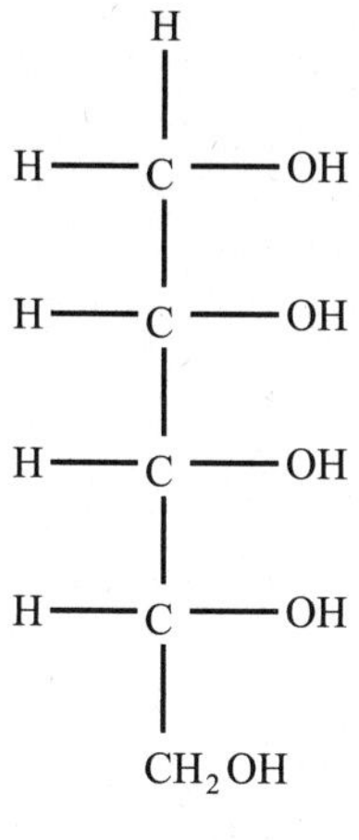

Figure 7-10: D-ribitol.

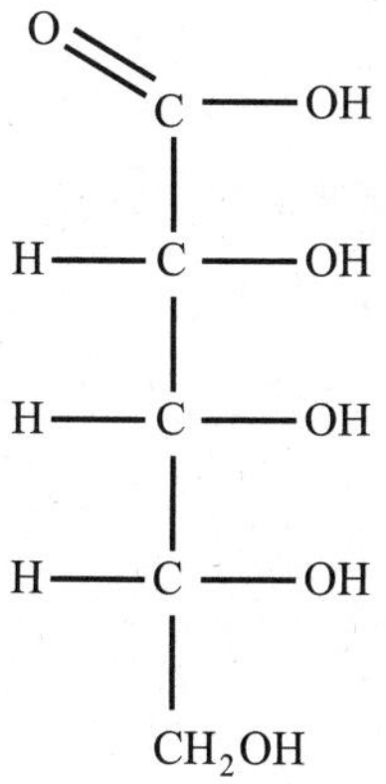

$$
\begin{array}{c}
\text{H} \\
| \\
\text{H}\!-\!\text{C}\!-\!\text{OH} \\
| \\
\text{H}\!-\!\text{C}\!-\!\text{OH} \\
| \\
\text{H}\!-\!\text{C}\!-\!\text{OH} \\
| \\
\text{H}\!-\!\text{C}\!-\!\text{OH} \\
| \\
\text{CH}_2\text{OH}
\end{array}
$$

Figure 7-11: D-ribonic acid, an aldonic acid.

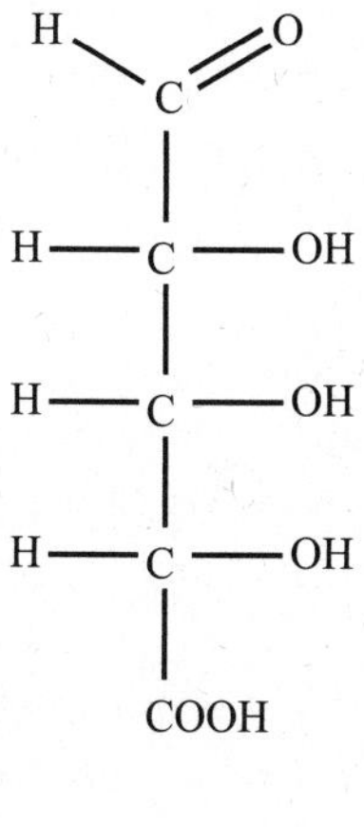

$$
\begin{array}{c}
\text{O} \\
\parallel \\
\text{C}\!-\!\text{OH} \\
| \\
\text{H}\!-\!\text{C}\!-\!\text{OH} \\
| \\
\text{H}\!-\!\text{C}\!-\!\text{OH} \\
| \\
\text{H}\!-\!\text{C}\!-\!\text{OH} \\
| \\
\text{CH}_2\text{OH}
\end{array}
$$

Figure 7-12: D-ribouronic acid, a uronic acid.

$$
\begin{array}{c}
\text{H}\!-\!\text{C}\!=\!\text{O} \\
| \\
\text{H}\!-\!\text{C}\!-\!\text{OH} \\
| \\
\text{H}\!-\!\text{C}\!-\!\text{OH} \\
| \\
\text{H}\!-\!\text{C}\!-\!\text{OH} \\
| \\
\text{COOH}
\end{array}
$$

Monosaccharides, like all alcohols, may react with acids to form *esters*. The combination with phosphoric acid (phosphate sugar) is a biologically important reaction. Any of the alcohol groups may react. Figure 7-13 shows one example: *D-ribose-1-phosphate*. (The *1* refers to the attachment of the phosphate group to C_1.)

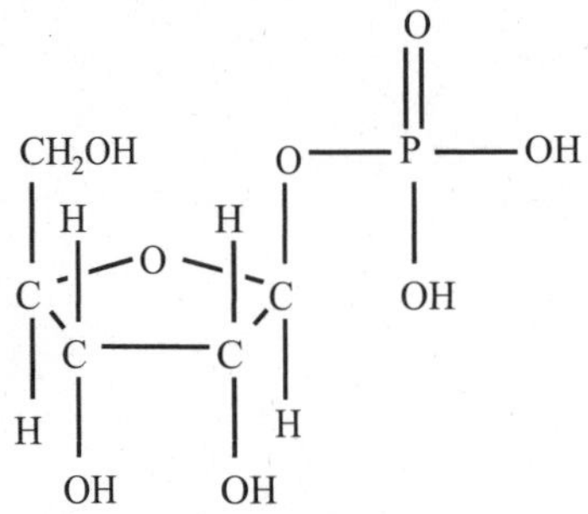

Figure 7-13: D-ribose-1-phosphate.

The most common monosaccharides

Glucose, or blood sugar, is also known as *dextrose*. The anomeric carbon is part of a hemiacetal, and the name of the pyranose structure is *glucopyranose*.

Blood is commonly tested for blood glucose levels. These levels are controlled by the hormone *insulin*, which is produced within the body in the pancreas. In a healthy human, blood glucose levels rise slightly after eating. The pancreas then releases insulin to keep the levels from rising too high. A healthy individual has a fasting blood sugar of 70 to 99 milligrams of glucose per deciliter of blood, and 70 to 145 milligrams per deciliter two hours after eating. The American Diabetes Association associates blood glucose levels of 126 milligrams per deciliter (fasting) or 200 milligrams per deciliter (two hours after eating) with *diabetes* — the inability of the pancreas to produce enough insulin.

The simplest aldose is *glyceraldehyde*, and the simplest ketose is *dihydroxyacetone*. Figure 7-14 shows the structures of these two compounds.

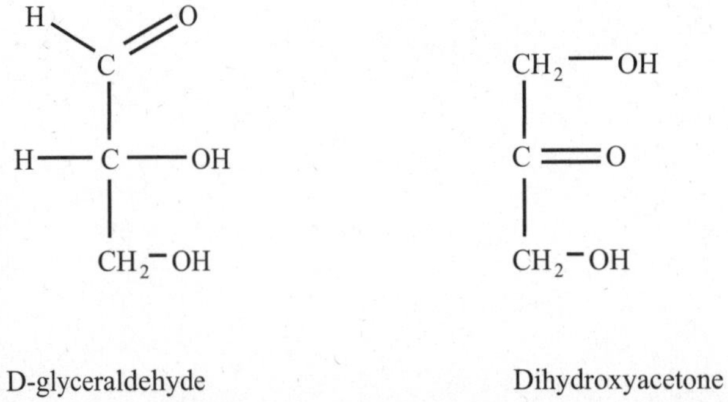

Figure 7-14: Glyceraldehyde and dihydroxyacetone.

D-glyceraldehyde

Dihydroxyacetone

The beginning of life: Ribose and deoxyribose

The monosaccharides *D-ribose* and *D-deoxyribose* are important components of the nucleic acids. They're present in these complex molecules in the form of a furanose ring. In addition, they're present as the β anomer. The difference between these two monosaccharides is that deoxyribose has one less oxygen atom, hence the *deoxy*. The "missing" oxygen atom is at C_2. The structures of these two sugars appear in Figure 7-15.

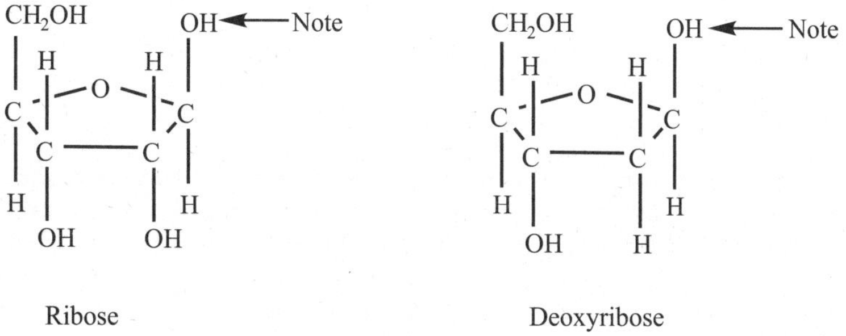

Figure 7-15: The arrows point to the positions of the alcohol groups leading to these becoming the β anomers.

Ribose

Deoxyribose

Sugars Joining Hands: Oligosaccharides

The joining of two or more monosaccharides forms either an *oligosaccharide,* with two to ten monosaccharide units, or a *polysaccharide,* a polymer having many more monosaccharide units. One or more glycoside linkages hold the monosaccharides together. The simplest, and most common, oligosaccharides are the disaccharides.

Keeping it simple: Disaccharides

A *disaccharide* is an oligosaccharide composed of two monosaccharide units. The best-known disaccharide (and surely the tastiest) is probably *sucrose,* which you know as table sugar or cane sugar. Each molecule of this sugar is a combination of a glucose molecule and a fructose molecule. Many other important disaccharides exist, among them *maltose* (malt sugar) and *cellobiose,* each of which contains two molecules of glucose. Because of maltose's simplicity, where two identical monosaccharides are joined, we use maltose to illustrate several points concerning disaccharides, and, by implication, other oligosaccharides and polysaccharides. The structure of maltose appears in Figure 7-16.

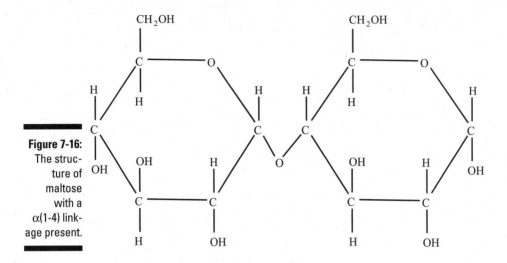

Figure 7-16:
The structure of maltose with a α(1-4) linkage present.

The oxygen atom joining the two glucose rings of the maltose molecule in Figure 7-16 is a *glycoside linkage* — a β(1-4) linkage. The α refers to the anomeric form of the ring on the left. If β-D-glucose were present instead, then *cellobiose* would result (see Figure 7-17). The 1-4 indicates that C_1 of the left ring links to C_4 of the right ring. The loss of a water molecule accompanies the formation of the linkage, which locks the left ring so that mutarotation is no longer possible. The locked ring is also no longer a reducing sugar. But mutarotation can still occur on the right ring.

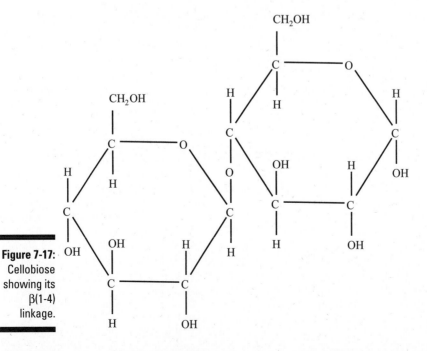

Figure 7-17:
Cellobiose showing its β(1-4) linkage.

Sucrose is a disaccharide, like maltose. It forms when D-glucose links to a D-fructose by a α(1-2) linkage. This situation locks both rings so that mutarotation of neither ring can occur. The formation of sucrose appears in Figure 7-18.

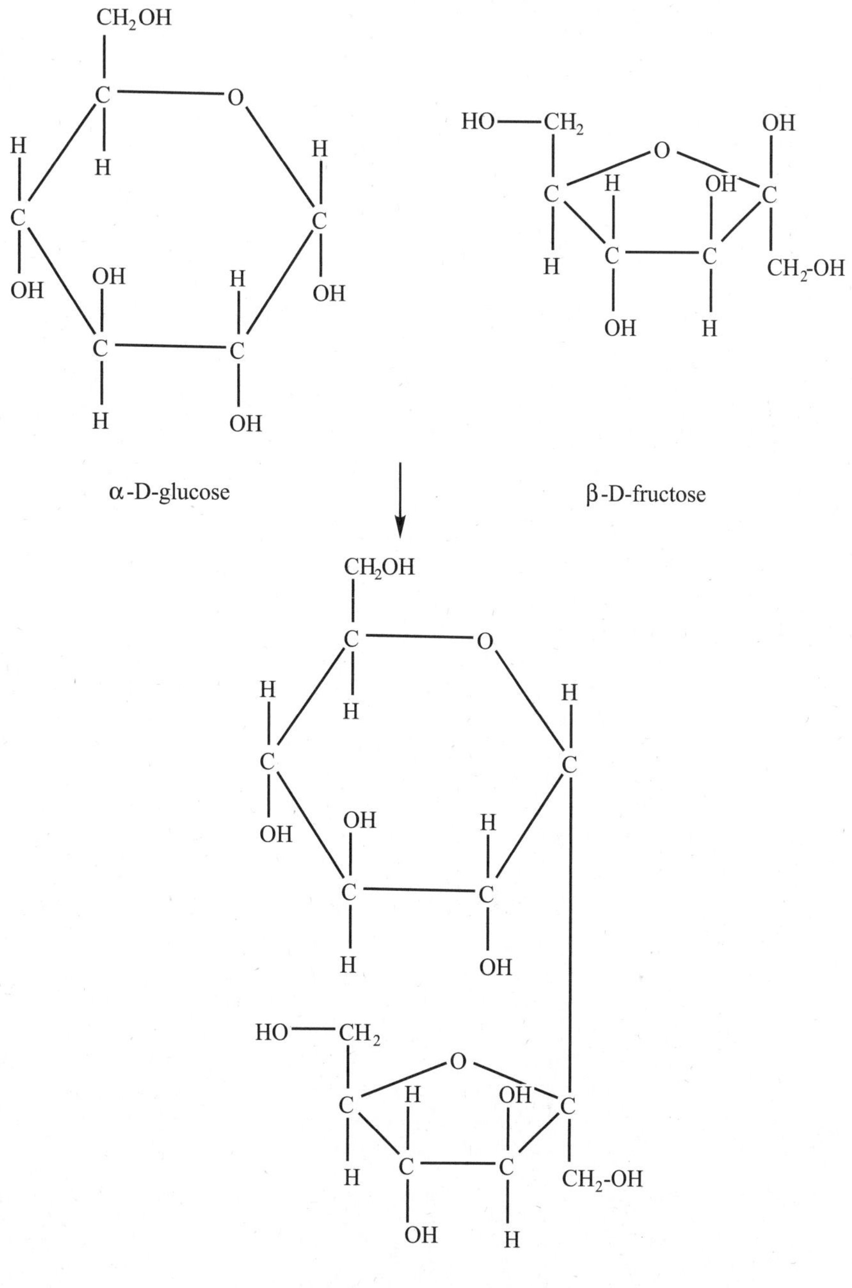

α-D-glucose

β-D-fructose

Figure 7-18:
Structure
of sucrose,
formed
by joining
αD-glucose
and β-D-
fructose.

Sucrose

If we had a scale that measured sweetness level and we assigned sucrose a value of 100, then the sweetness level of glucose would be 74 and that of fructose would be 173. Fructose, found in corn syrup, is the sweetest common sugar, meaning you need less of it to make foods taste sweet. Less sugar translates to fewer calories. Some other naturally occurring, sweet-tasting proteins are hundreds of times sweeter than sugar.

Quite a few artificial sweeteners are used in commercial products. The best known are *saccharin* (about 500 times as sweet as sucrose), *aspartame* (200 times as sweet as sucrose), and *sucralose* (marketed as *Splenda*), which is a whopping 600 times as sweet as sucrose. Sucralose is created by replacing three of the hydroxyl groups of sucrose with chlorines.

Starch and cellulose: Polysaccharides

The two most important polysaccharides are *starch* and *cellulose.* Both of these are polymers of D-glucose. The basic difference between these two polymers is the linkages between the glucose units. Starch is related to maltose and cellulose is related to cellobiose.

Bread, pasta, and potatoes: Starches

Of all the carbohydrates, we think starches are our favorite. Bring on the potatoes and pastas! The different types of these lovely, delicious polysaccharides are very closely related by the linkages between their monomer units. Starch is a polymer of α-D-glucose. Starch comes in three common types: *amylose, amylopectin,* and *glycogen.* Amylase is the combination of α(1-4) glucose groups. Amylopectin, like amylase, has α(1-4) glucose linkages, but in addition, it has α(1-6) branches. Glycogen, or animal starch, is similar to amylopectin except that it has more branches. All three are useful in storing glucose, and all three show an intense dark blue color in the presence of iodine — a simple and useful test.

Keeping the termites happy: Cellulose

Ever wonder why you can eat a potato but not a tree? Cellulose is very similar to starch except that the linkages are β(1-4) glucose. (These linkages are shown in Figures 7-16 and 7-17.) The primary use of cellulose in nature is structure. Cleavage of the linkages is only possible with enzymes produced by certain bacteria or fungi. For this reason, only certain creatures — such as termites and ruminants like cows, which harbor particular bacteria in their gastrointestinal tracts — can digest and utilize cellulose as an energy source. Cellulose is one of the most abundant biochemicals on earth.

Biological connective tissue: Acidic polysaccharides

One of the major uses of polysaccharides in the body is the area of connective tissues, the compounds that hold human parts together. This group of tissues includes tendons, ligaments, and collagen. (Fuller lips, anyone?) Acidic polysaccharides are important to the structure and function of connective tissue. The repeating units of these polysaccharide derivatives are disaccharides. One of the components of the disaccharide is an *amino sugar* (where an amino group substitutes for an alcohol group). One or both of the components of the disaccharide unit have a negatively charged group (either a sulfate or a carboxylate). Examples are *hyaluronic acid* and *heparin.* The hyaluronate and heparin repeating units appear in Figure 7-19.

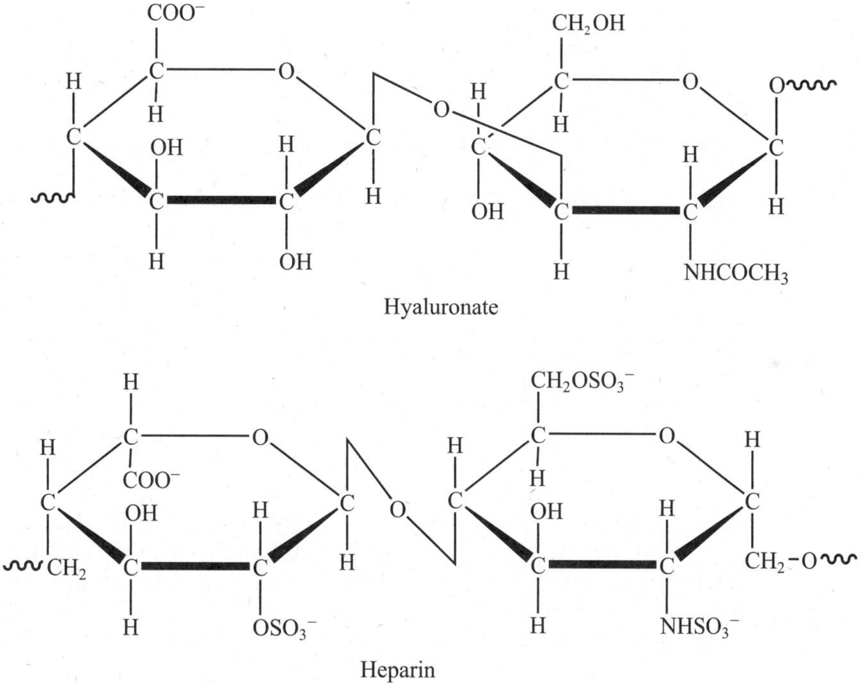

Hyaluronate

Figure 7-19: Disaccharide repeating units in hyaluronate and heparin.

Heparin

Heparin is used to treat and prevent blood clots from forming, especially in the lungs and legs. It's commonly used after dialysis, after surgery, or when a patient has been unable to move for extended periods of time. It acts as an anticoagulant by binding to one of the anticlotting proteins, increasing its efficiency up to a thousand-fold.

Glycoproteins

Most of the proteins occurring in blood serum are *glycoproteins,* which are proteins with carbohydrates attached. The presence of the carbohydrate tends to increase the hydrophilic nature of the protein. In general, the linkage is by attachment to an asparagine, serine, or threonine residue. Some soluble proteins and some membrane proteins are glycoproteins (such as some protein hormones and their membrane receptors). We discuss glycoproteins again at various times in later chapters.

The Aldose Family of Sugars

Glucose and ribose are two important aldose sugars that appear in many places in this book. However, there are a large number of additional aldose sugars. An important subgroup of these aldose sugars is the D-family. The simplest of these is triose (3-carbon) glyceraldehyde. There are two tetroses (4-carbon), four pentoses (5-carbon), and eight hexoses (6-carbon).

Figure 7-20 shows the structures of all members of the D-aldose family. Each of these aldose sugars has an enantiomer. The mirror image of each aldose in the figure is the L-enantiomer. However, the L-enantiomers aren't biologically important.

The representations in Figure 7-20 are Fischer projections. At the top of each projection is a circle. This circle, or "head," represents the aldehyde group present in all the aldoses. The aldehyde carbon is number 1, and the other carbon atoms are numbered from it. The last carbon atom, at the very bottom, is in a $-CH_2OH$ group. Neither the top nor bottom carbon atom is chiral, but all the intermediate carbon atoms are chiral. Each of the remaining carbon atoms has an –H and an –OH attached. The short line, to the left or right, indicates the –OH group. For all D-aldose sugars, the –OH on the highest numbered chiral carbon (the next-to-the-last carbon atom in the chain) is on the right. (For an L-aldose, this –OH would be on the left.) The name of each aldose appears below its structure.

You can remember the members of the D-aldose family in several ways. In all cases, the bottom chiral carbon is D (–OH to the right). The far right structure in each row has all –OH groups going to the right. The next structures to the left begin at the top, switch one –OH to the left, and move down to the next-to-last chiral carbon (always to the right). Figure 7-21 indicates the pattern for the bottom row of Figure 7-20. Figure 7-22 gives the overall pattern in Figure 7-20.

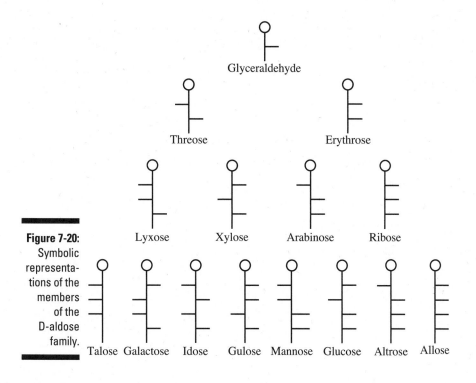

Figure 7-20: Symbolic representations of the members of the D-aldose family.

Glyceraldehyde

Threose   Erythrose

Lyxose   Xylose   Arabinose   Ribose

Talose   Galactose   Idose   Gulose   Mannose   Glucose   Altrose   Allose

Figure 7-21: The relative positions of the –OH groups in the bottom row of Figure 7-20. (+) indicates right and (–) indicates left.

_	+	_	+	_	+	_	+
_	–	+	+	–	–	+	+
_	–	–	–	+	+	+	+
+	+	+	+	+	+	+	+
T	G	I	G	M	G	A	A

Figure 7-22: The overall pattern in Figure 7-20.

				G				
	T					E		
L			X		A		R	
T	G	I	G	M	G	A	A

The following mnemonic gives the bottom row in Figure 7-20 from right to left: "ALL ALTruists GLadly MAke GUmbo In GALlon TAnks."

Here's how to construct the bottom row of Figure 7-20:

1. **Start the eight Fischer projections (–CHO at the top and –CH$_2$OH at the bottom).**

2. **C$_5$ (above –CH$_2$OH) has –OH on the right.**

3. **C$_4$ –OH groups for the first four are on the left and for the next four are on the right.**

4. **C$_3$ –OH groups are two left, two right, two left, and two right.**

5. **C$_2$ –OH groups are left, right, left, right, left, right, left, and right.**

If all else fails, use the mnemonic!

Chapter 8

Lipids and Membranes

. . . .

. . . .

Along with cholesterol, lipids tend to have a bad reputation in today's world, even though they're absolutely necessary to good health. The *lipids* are an exceedingly diverse group of biologically important materials that are distinguished by solubility. A lipid is a member of a group of compounds that aren't soluble (or are only sparingly soluble) in water but are soluble in nonpolar solvents or solvents of low polarity. Lipids have a nonpolar nature because a large portion of the molecule contains only carbon and hydrogen. If the structure had significant amounts of oxygen or nitrogen, the substance would be more polar and hence more soluble in water.

Lovely Lipids: An Overview

Lipids have many important biological roles, including being highly concentrated energy sources, membrane components, and molecular signals. The diagram in Figure 8-1 shows the relationship among many of the different categories of lipids. Arachidonic acid, a fatty acid, appears in Figure 8-1 twice — once as the *precursor* (compound leading) to leukotrienes and prostaglandins and again as a member of the fatty acid group. We double-list arachidonic acid this way because of its very different roles in these two chemical pathways. (See the section "Prostaglandins, Thromboxanes, and Leukotrienes: Mopping Up" later in the chapter.)

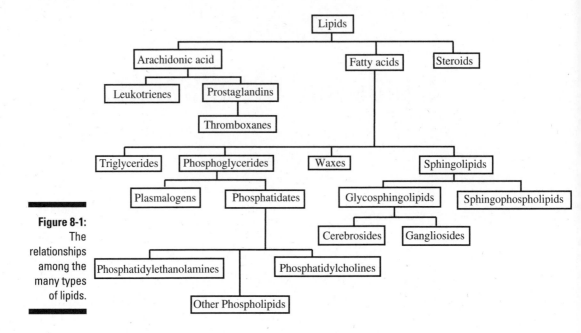

Figure 8-1:
The relationships among the many types of lipids.

Behavior of lipids

In the body, lipids provide energy storage and structure (as in cell membranes) and regulate bodily functions. Many of the lipids work like soaps and detergents. Like soaps, lipids have a nonpolar region — usually a fatty acid — and a polar region. Figure 8-2 shows a representation of a soap's structure.

Figure 8-2:
Representation of a soap.

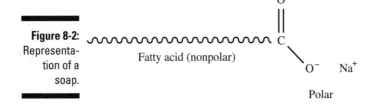

In water, soap forms a *micelle* (see Chapter 2) in which the nonpolar portions of the different molecules coalesce and leave the polar portions on the outside, next to the water. If any other nonpolar material is present, such as grease from dirty dishes, it tends to migrate to the micelle's interior. With the polar portions of the soap molecules on the outside, the micelle appears as one large polar molecule instead of a number of smaller molecules that have polar and nonpolar regions. See how knowing a little chemistry makes washing dishes a little more fun?

The *dual solubility* nature of soap is why it removes grease or oil from your skin or clothes. The grease or oil is nonpolar and, therefore, isn't soluble in water. The soap forms a micelle that surrounds the grease/oil in the micelle's nonpolar portion. The soap micelle's polar end is soluble in water, allowing the grease and oil to be removed during rinsing.

Although many different kinds of lipids exist, the discussion in this chapter focuses on these four types:

- ✔ **Fatty acids and derivatives (esters):** Fats, oils, and waxes
- ✔ **Complex lipids:** Phosphoglycerides and sphingolipids
- ✔ **Steroids**
- ✔ **Arachidonic acid derivatives:** Prostaglandins, thromboxanes, and leukotrienes

Fatty acids in lipids

Lipids are important not only as individual molecules but also in terms of their interactions with other lipids and nonlipids in the formation of lipid bilayers or cell membranes. These interactions occur both at the cell boundary and around some interior structures. The fatty acid portions of the lipids are especially important in their physical and chemical properties. The naturally occurring fatty acids have a few key features:

- ✔ They're all straight-chained and typically have 10 to 20 (but sometimes more) carbon atoms.
- ✔ They have an even number of carbon atoms.
- ✔ If carbon-carbon double bonds are present, only the cis-isomer is present.

Table 8-1 lists a few of the common fatty acids.

Table 8-1	Common Fatty Acids
Chemical Name	*Structural Formula*
Lauric acid	$CH_3(CH_2)_{10}COOH$
Myristic acid	$CH_3(CH_2)_{12}COOH$
Palmitic acid	$CH_3(CH_2)_{14}COOH$
Palmitoleic acid	$CH_3(CH_2)_5CH=CH(CH_2)_7COOH$
Stearic acid	$CH_3(CH_2)_{16}COOH$
Oleic acid	$CH_3(CH_2)_7CH=CH(CH_2)_7COOH$
Linoleic acid	$CH_3(CH_2)_3(CH_2CH=CH)_2(CH_2)_7COOH$
Linolenic acid	$CH_3(CH_2CH=CH)_3(CH_2)_7COOH$
Arachidonic acid	$CH_3(CH_2)_4(CH=CHCH_2)_4(CH_2)_2COOH$

A *wax* is a simple ester of a fatty acid and a long-chain alcohol. The fatty acid typically contains at least 10 carbon atoms, whereas the alcohol portion is typically 16 to 30 carbon atoms. In general, a wax, such as the wax in your ears, serves as a protective coating. In the plant world, waxes offer protection from pests and water loss. Because waxes tend to be somewhat unreactive (thank goodness for our ears' sake), we don't discuss waxes in much detail in this book.

A Fatty Subject: Triglycerides

Fats (and oils) are *triglycerides* or *triacylglycerols*. That is, they're triesters of fatty acids with glycerol. *Glycerol* is a trihydroxy alcohol (see Figure 8-3). In a fat, each of the three alcohol groups becomes part of an ester through the reaction with a fatty acid. The fatty acids may or may not be the same within the triglyceride.

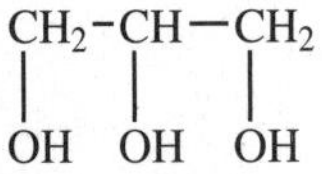

Figure 8-3:
Structure of
glycerol.

$$CH_2-CH-CH_2$$
$$OH \quad OH \quad OH$$
Glycerol

Properties and structures of fats

The practical difference between a fat and an oil is that a fat is a solid at room temperature whereas an oil is a liquid. That said, two important structural factors distinguish a fat from an oil: the size of the fatty acids and the presence or absence of double bonds. The longer the fatty acid chain, the higher the melting point. The greater the number of carbon-carbon double bonds, the lower the melting point.

A *saturated* fat consists of fatty acids with no carbon-carbon double bonds. An *unsaturated* fat has a carbon-carbon double bond, and a *polyunsaturated* fat has multiple double bonds.

More than 70 known naturally occurring fatty acids exist. In most natural fats, the double bonds have a cis geometry. The presence of double bonds puts "kinks" in the carbon chain, which prevent the fatty acid chains from stacking together as roughly parallel chains. The inability of unsaturated fatty acid chains to stack together inhibits the fat's ability to solidify at room temperature. Think about that takeout container of Mexican food. After it sits on the counter for a while, the oil may solidify, indicating that there are more saturated fats present. This isn't true of the liquid olive oil coauthor John keeps next to his stove.

The treatment of an unsaturated fat or oil with hydrogen in the presence of a catalyst such as nickel leads to hydrogenation of some or all of the carbon-carbon double bonds, forming carbon-to-carbon single bonds. This procedure changes an unsaturated fat into a saturated fat. In most cases, only partial hydrogenation takes place, and the hydrogenation raises the compound's melting point. By this procedure, converting an oil (liquid) into a fat (solid) is possible. Incomplete hydrogenation may change some of the cis arrangements into trans arrangements, producing a *trans fat*. (You may have read about the effects of trans fats in lowering HDL, the "good" cholesterol, and raising LDL, the "bad" cholesterol. Wow, two for the price of one.) This is the way that vegetable oil is converted into margarine.

Figure 8-4 shows the structure of a typical fat. Note that the two upper fatty acid chains (saturated) "stack" next to each other, but the lower chain (unsaturated) does not.

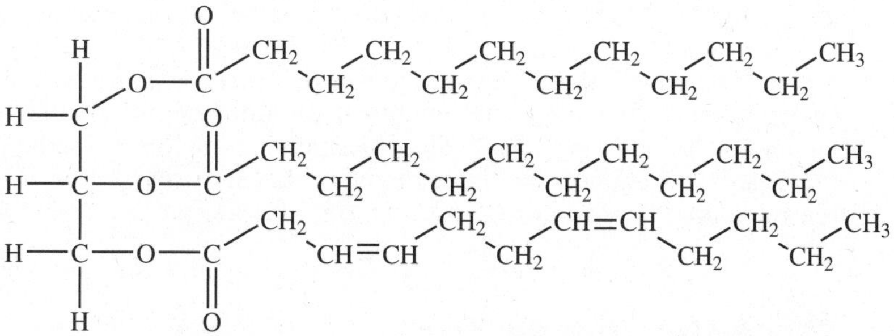

Figure 8-4:
Structure of a typical fat: Upper chains are saturated; bottom chain is unsaturated.

Cleaning up: Breaking down a triglyceride

For centuries, the treatment of a fat (commonly animal fat) with a strong base catalyst (generally lye, or sodium hydroxide) has been used to produce soap. John's grandmother made soap by boiling hog fat with wood ashes, which contain potassium hydroxide and sodium hydroxide. She then skimmed off the soap and pressed it into cakes. Unfortunately, Granny wasn't very good at getting all the proportions just right and tended to use too much base, making the soap very alkaline. In this kind of reaction, called a *saponification reaction,* hydrolysis of the ester groups in the presence of a base yields glycerol and the carboxylate ions of the three fatty acids. A soap is really a sodium or potassium salt of a fatty acid. The calcium and magnesium analogues, on the other hand, are insoluble. If the soap is used with *hard* water (containing calcium or magnesium ions), it precipitates as a greasy scum: bathtub ring.

Acids also catalyze the hydrolysis of a fat to produce a glycerol and three fatty acids. Acid hydrolysis is reversible, whereas the presence of excess base inhibits the reverse of saponification. During digestion, lipases break down triglycerides, and bile salts make the fatty acid portions soluble. A *lipase* is an enzyme that catalyzes the decomposition of a fat. *Bile salts* are oxidation products of cholesterol that act as detergents to make the products of the breakdown soluble. In humans, absorption of the products occurs in the small intestine.

No Simpletons Here: Complex Lipids

So far, we've been discussing simple lipids. However, some lipids are somewhat more complex. (You didn't really think we were going to keep it simple, did you?) In general, complex lipids are esters of glycerol or some other alcohol. The two major categories of complex lipids are the phosphoglycerides and the sphingolipids. The *phosphoglycerides* are the plasmalogens and the phosphatidates. The *sphingolipids* are the glycosphingolipids and the sphingophospholipids. (Further subdivision is shown back in Figure 8-1.)

A *phospholipid* is either a phosphoglyceride or a sphingophospholipid. Phospholipids are major components of membranes. Any carbohydrate-containing lipid is a *glycolipid.* The classifications of lipids overlap. (As you may have noticed, nothing in biochemistry is ever truly simple.) For this reason, a lipid may fall into more than one subcategory.

Phosphoglycerides

The alcohol here is glycerol, to which two fatty acids and a phosphoric acid are attached as esters. This basic structure is a phosphatidate — an important intermediate in the synthesis of many phosphoglycerides. The presence of an additional group attached to the phosphate allows for many different phospho-glycerides, like scaffolding, which can be used to build a wide variety of different looking buildings.

By convention, these compounds' structures show the three glycerol carbon atoms vertically with the phosphate attached to carbon atom number 3 (at the bottom). The occurrence of phosphoglycerides is almost exclusive to plant and animal cell membranes. Plasmalogens and phosphatidates are examples. These are also known as *glycerophospholipids.*

Plasmalogens

Plasmalogens are a type of phosphoglyceride. The first carbon of glycerol has a hydrocarbon chain attached via an ether, not ester, linkage. Ether linkages are more resistant to chemical attack than ester linkages. The second (central) carbon atom has a fatty acid linked by an ester. The third carbon most commonly links to an ethanolamine or choline by means of a phosphate ester. These compounds are key components of the membranes of muscles and nerves.

Phosphatidates

Phosphatidates are lipids in which the first two carbon atoms of the glycerol are fatty acid esters, and the third is a phosphate ester. The phosphate serves as a link to another alcohol — usually ethanolamine, choline, serine, or a carbohydrate. The identity of the alcohol determines the subcategory of the phosphatidate. The phosphate has a negative charge and, in the case of choline or serine, a positive quaternary ammonium ion. (Serine also has a negative carboxylate group.) The presence of charges gives a "head" with an overall charge. The phosphate ester portion ("head") is hydrophilic (dissolves in water), whereas the remainder of the molecule, the fatty acid "tail," is hydrophobic (not attracted or soluble in water). These are important components for the formation of lipid bilayers that are found in all of our cells.

Phosphatidylethanolamines, phosphatidylcholines, and other phospholipids are examples of phosphatidates. Figure 8-5 illustrates examples of a phosphatidylethanolamine and a phosphatidylcholine.

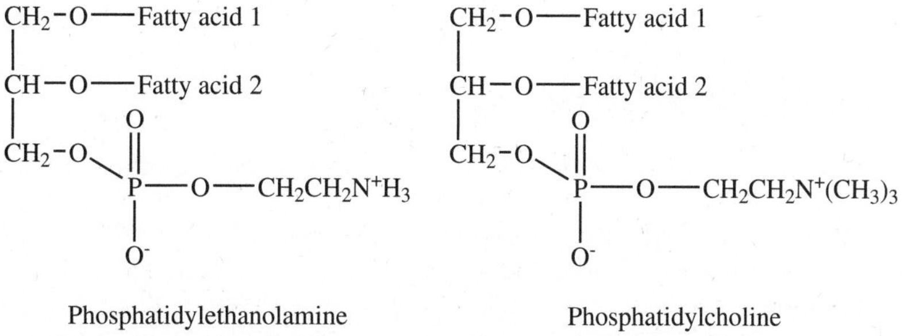

Figure 8-5: Examples of the general formulas of a phosphatidyl-ethanolamine and a phosphatidyl-choline.

The structures of some of the alcohols present in lipids appear in Figure 8-6.

$HO\text{-}CH_2CH_2\text{-}N^+(CH_3)_3$

Choline

$$CH_2\text{-}CH\text{-}CH_2$$
$$|\quad\ |\quad\ |$$
$$OH\ \ OH\ \ OH$$

Glycerol

Figure 8-6:
Alcohol components of lipids.

$HOCH_2CH_2NH_2$

Ethanolamine

$HOCH_2CH\text{-}COOH$
$\qquad\qquad |$
$\qquad\quad NH_2$

Serine

Phosphatidylethanolamines

Phosphatidylethanolamines are the most common phosphoglycerides in animals and plants. In animals, many of these are the *cephalins,* which are present in nerves and brain tissue. They're also factors involved in blood clotting. Recall that the phosphate has a negative charge and that the nitrogen of the enthanolamine is a quaternary ammonium ion with a positive charge.

Phosphatidylcholines

Phosphatidylcholines are the lecithins. Choline is the alcohol, with a positively charged quaternary ammonium ion, bound to the phosphate, with a negative charge. Lecithins are present in all living organisms. An egg yolk has a high concentration of lecithins, which are commercially important as an emulsifying agent in products such as mayonnaise. Lecithins are also present in brain and nerve tissue and serve as an emulsifier in the lungs.

Other phospholipids

Many other phospholipids exist, some of which are glycolipids. The glycolipids include phosphatidyl sugars where the alcohol functional group is part of a carbohydrate. Phosphatidyl sugars are present in plants and certain microorganisms. A carbohydrate is very hydrophilic because of the large number of hydroxyl groups present.

Sphingolipids

Sphingolipids occur in plants and animals and are especially abundant in brain and nerve tissue. In these lipids, *sphingosine* (see Figure 8-7) replaces glycerol. The alcohol groups in the sphingosine may form esters, just like the similar groups on glycerol. The amino group can form an amide. The combination of a fatty acid and sphingosine, via an amide linkage, is a *ceramide,* which is an intermediate in the formation of other sphingolipids.

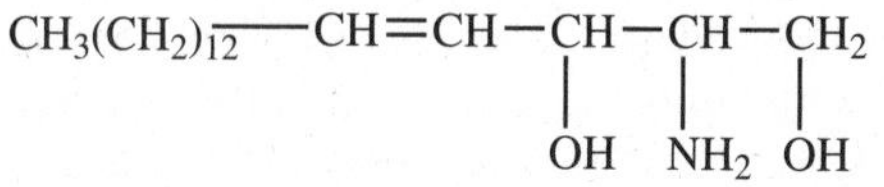

Figure 8-7:
Structure of
sphingosine.

$$CH_3(CH_2)_{\overline{12}}\!\!-\!\!CH\!\!=\!\!CH\!\!-\!\!CH\!\!-\!\!CH\!\!-\!\!CH_2$$
$$\phantom{CH_3(CH_2)_{12}\!\!-\!\!CH\!\!=\!\!CH\!\!-\!\!}OH\quad NH_2\quad OH$$

Glycosphingolipids and cerebrosides

A *glycosphingolipid* is an important membrane lipid containing a carbohydrate attached to a ceramide. The carbohydrate serves as a polar (hydrophilic) head. The carbohydrate may be either a monosaccharide or an oligosaccharide. The carbohydrate sequence in the oligosaccharide is important in helping these compounds recognize other compounds in biochemical reaction sequences. The carbohydrate portion is always on the outside of the membrane.

A *cerebroside* consists of a monosaccharide attached to a ceramide. The carbohydrate is either glucose or galactose. Cerebrosides are present in nerve and brain cells, though most animal cells contain some of these compounds.

Gangliosides

Gangliosides are sphingolipids with complex structures. The ceramide has an oligosaccharide attached that contains three to eight monosaccharide units, which may or may not be substituted. Gangliosides are very common as part of nerve cells' outer membranes, where the sugar sequence leads to cell recognition and communication. Small quantities of gangliosides are part of the outer membranes of other cells. When present in a membrane, the carbohydrate portion is always extracellular.

Sphingophospholipids

Sphingophospholipids contain sphingosine, a fatty acid, phosphate, and choline. An example is *sphingomyelin,* an important constituent of the myelin sheath surrounding the axon of all nerve cells. Multiple sclerosis, among other diseases, is a consequence of a fault with the myelin sheath. Sphingomyelin is the most common of the sphingolipids, and it's the only sphingosine phospholipid found in membranes.

Membranes: The Bipolar and the Bilayer

One use of lipids is in the construction of membranes. *Membranes* separate regions both in and around cells. A typical membrane, as shown in Figure 8-8, is a lipid bilayer or bimolecular sheet. The lipids' polar portions, or heads, are on the bilayer's outside edges, whereas the nonpolar portions, the tails, are in the interior. The heads appear as circles in our illustrations, and the tails appear as strings. The tails are usually long fatty acid chains. The hydrophilic heads (mentioned in the "Phosphatidates" section earlier in this chapter), often with a charge, are in contact with aqueous material, and the hydrophobic tails are away from the aqueous material. Interactions between the hydrophobic tails are the key factors leading to the formation of lipid bilayers. *Lipid bilayers* tend to form closed structures or compartments to avoid having exposed hydrophobic edges. The membranes tend to be self-sealing. We've mentioned lipid bilayers previously; now we are going to look at them in more depth.

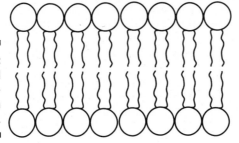

Figure 8-8:
A simplified representation of a lipid bilayer.

Actual cell membranes aren't as symmetrical as the one shown in Figure 8-8. This asymmetry is due in part to the presence of other components and in part to differences between the intracellular and extracellular surfaces. If the fatty acid portions aren't saturated, the tails don't form parallel structures, and "holes" are present within the bilayer. These holes are an essential feature leading to membrane fluidity. Other components include proteins and cholesterol. The carbohydrate portion of glycolipids is on the extracellular side of the bilayer instead of the intracellular side.

Salmon exhibit this change in membrane fluidity as they move to and from different water temperatures. Because salmon are coldblooded, their bodies adjust the amount of cholesterol in their cells when moving between different water temperatures to maintain the correct flexibility of their bodies, not too stiff, not too limber.

Polar materials can't readily pass through the hydrophobic region of membranes, and nonpolar materials can't readily pass through the hydrophilic outer region. Because of their small size and high concentration, water molecules can transverse the bilayer faster than ions and most other polar molecules. In actual cells, certain mechanisms allow material to cross the bilayer but require other components to be present in the bilayer. These components, mostly proteins, give selective permeability to the membranes. In addition, other materials, such as cholesterol, are necessary to serve other functions, such as stiffening the membrane.

Membranes may contain roughly 20 to 80 percent protein, which may be *peripheral* (on the membrane's surface) or *integral* (extending into or through the membrane). Integral proteins interact extensively with the bilayer's hydrophobic portion, as illustrated in Figures 8-9 and 8-10.

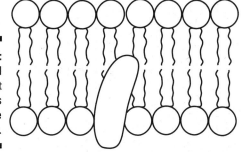

Figure 8-9:
An integral protein that doesn't pass through the membrane.

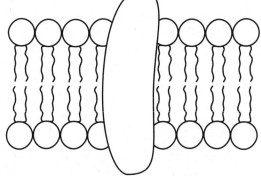

Figure 8-10:
An integral protein passing through the membrane.

Peripheral proteins typically bind to the surface through electrostatic or hydrogen bonding, although covalent interactions are possible. Proteins are important for most membrane processes. If the protein is a glycoprotein, the carbohydrate portion lies on the membrane's external side and is important to intercellular recognition.

Crossing the wall: Membrane transport

A lipid bilayer is, by its nature, impermeable to polar molecules and ions (hydrophilic species). Nevertheless, cells need to be able to bypass this feature and get hydrophilic materials in and out. Cells can circumvent impermeability in two ways: A *pump* involves active transport using energy to work against a concentration gradient, and a *channel* involves passive transport or facilitated diffusion using a concentration gradient.

Nonpolar molecules are *lipophilic* and dissolve in the lipid bilayer. In general, lipophilic materials pass through the membrane by simple diffusion along a concentration gradient. Pumps and channels exist mainly to allow hydrophilic species to transverse the bilayer's hydrophobic region.

Pumps

Pumps require energy to function. In many cases, the hydrolysis of ATP provides the needed energy. The generic name for this type of pump is a *P-type ATPase.* The name derives from the transfer of a phosphate from an ATP to an intermediate, a step that's essential to the pump's action. Pumps can transfer species other than ions.

Most animal cells have a high potassium ion and a low sodium ion concentration relative to the extracellular environment. Generating and maintaining this gradient requires energy. The transport system is the Na$^+$-K$^+$ pump, also referred to as *Na$^+$-K$^+$ ATPase.* Hydrolysis of ATP provides the energy to transport potassium ions into the cell and sodium ions out of the cell. Both the sodium and potassium ions must be simultaneously bound to the pump. The pump simultaneously transports three sodium ions out of the cell as it transports two potassium ions in.

Not all pumps require the hydrolysis of ATP to supply energy. Some utilize the transport of one species to facilitate the transport of another. The transport of one species with the concentration gradient can pump another against the concentration gradient. The responsible membrane proteins are the cotransporters or secondary transporters. *Cotransporters* may be either *symporters* or *antiporters.* In a symporter, both transported species move in the same direction, whereas in an antiporter, the species move in opposite directions. The sodium-calcium exchanger is an example of an antiporter that pumps three sodium ions into a cell for every two calcium ions pumped out. Some animal cells use a symporter to pump glucose coupled with sodium ions into the cells. Wow, take a couple of deep breaths and a sip of water and contemplate this amazingly efficient product of evolution before you go on!

Channels

A channel provides a means of passively transporting a species across a membrane. Transporting a species through a channel can be more than 1,000 times as fast as using a pump. A channel is technically a protein-based tube running through the membrane, but its behavior is significantly more complicated.

Channels are highly selective. Some select by size — sodium is smaller than potassium — whereas others differentiate between anions and cations. A channel exists in an open state to allow transport and in a closed state to inhibit it. Some type of regulation is required to convert a channel between an open and a closed state. When a chemical potential regulates the channel, it's a *voltage-regulated gate.* The regulation may be due to specific chemicals. Chemically controlled regulation is *ligand-gated.* After the appropriate regulator is removed, the open channels spontaneously close.

When is a solid a liquid? The fluid mosaic model

The lipid bilayer structure gives much insight into the structure of membranes but little information about their function. Many functions of the membrane depend on its fluidity, best described by using the *fluid mosaic model.* In this model, the membrane serves as a permeability barrier and as a solvent for the integral proteins. Diffusion of membrane components along the membrane's plane — *lateral diffusion* — is often rapid. In general, lipids move more rapidly than proteins, with some proteins being essentially immobile. Diffusion of membrane components across the membrane — *transverse diffusion* — is usually slow.

The membrane's fluidity depends on a number of factors. Bacteria adjust the fluidity by utilizing fatty acid chains; longer chains are less fluid than shorter chains. (Recall the salmon example used in the "Membranes: The Bipolar and the Bilayer" section.)The presence of double bonds makes the membrane more fluid. In animals, cholesterol controls the fluidity: The greater the cholesterol concentration, the less fluid the membrane. The transition from the rigid to the fluid state occurs at a temperature known as the *melting temperature, T_m.*

The best-known ligand-gated channel is the *acetylcholine receptor,* an important channel for the transmission of nerve impulses. When a nerve impulse reaches the junction between one nerve and the next — the *synapse* — it triggers the release of acetylcholine, which transverses the small gap to the next nerve and attaches to acetylcholine receptors. This attachment opens the channel, leading to inward sodium ion diffusion and outward potassium ion diffusion. The change in the ion concentrations transmits the nerve impulse into the second nerve cell.

The increase in the sodium ion concentration in the second nerve cell triggers a mechanism to remove sodium ions from the nerve cell. Later, another gate brings potassium ions back into the cell. This is called *electrochemical communication,* in which a signal is transmitted electrically and then chemically across a synapse, and then electrically again. This allows for rapid communication from one nerve cell to another.

Steroids: Pumping up

Steroids are another class of lipids. All steroids have the basic core shown in Figure 8-11. *A, B, C,* and *D* are common labels for the rings. Different steroids have additions to this basic structure; these may include side chains, other functional groups, and unsaturation or aromaticity of the rings.

Figure 8-11:
Basic
structure of
a steroid.

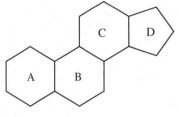

Cholesterol is the most abundant steroid precursor. It is a membrane component and serves as the source of many steroids and related materials. Cholesterol comes from the diet, but if insufficient cholesterol is available there, the body synthesizes it in the liver. The steroid hormones are chemical regulators/ligands produced from cholesterol.

Bile salts are a group of materials produced by the oxidation of cholesterol. Unlike cholesterol and the other lipids, bile salts are soluble in water. They're useful as "detergents" to aid in digestion.

The steroids you hear about in the news being used by athletes and bodybuilders are *anabolic steroids,* which increase the body's ability to prevent muscle breakdown and to produce muscle. They have structures similar to testosterone, whose function is to enhance male characteristics such as facial hair and muscle mass. However, steroids in large doses have serious side effects: impotence, reduced testicle size, liver tumors, enlargement of the heart, enlargement of the breasts in men, aggressive behavior, and so on. (Sounds great, doesn't it?) Their use without a valid prescription has been illegal since 1991.

Prostaglandins, Thromboxanes, and Leukotrienes: Mopping Up

Arachidonic acid — a 20-carbon, polyunsaturated fatty acid — serves as the direct or indirect starting material for the formation of *prostaglandins, thromboxanes,* and *leukotrienes.* Cells synthesize both leukotrienes and prostaglandins from arachidonic acid. Additional prostaglandins and thromboxanes come from the prostaglandin derived from arachidonic acid. All three classes of compounds are local hormones. Unlike other hormones, they're not transported via the bloodstream. They're short-lived molecules that alter the activity of the cell that produces them as well as neighboring cells.

All these compounds are extremely potent chemicals that serve as hormonal mediators. They also have many other medical applications and can cause medical problems. They're also known as *eicosanoids* — from the Greek for *twenty,* which alludes to the presence of 20 carbon atoms (see Figure 8-12).

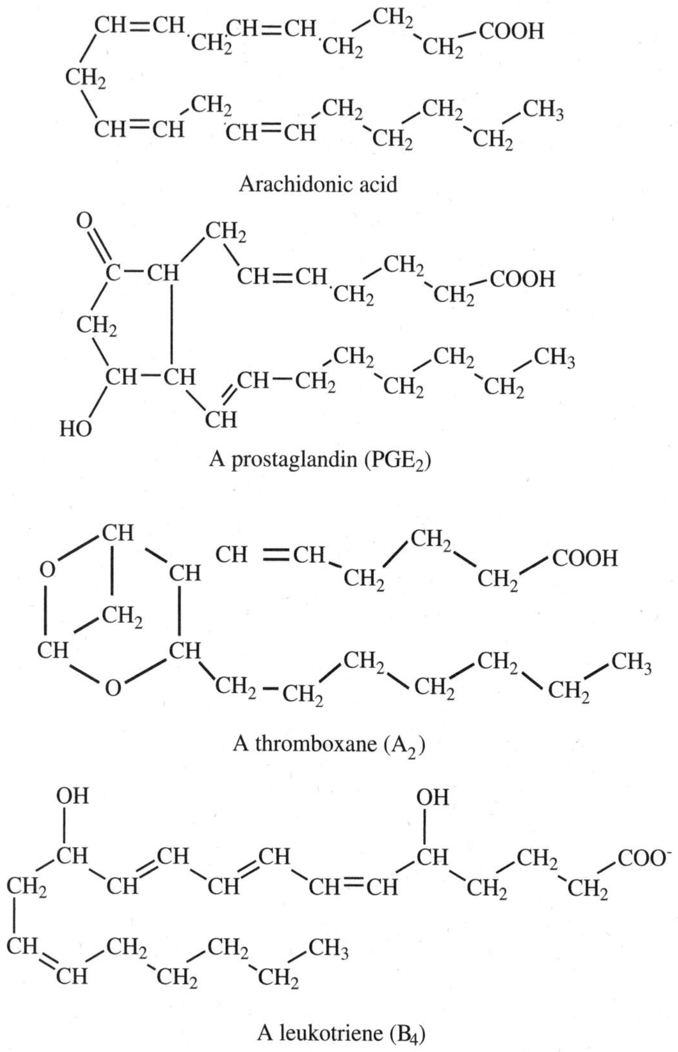

Figure 8-12:
Structures
of
arachidonic
acid and
a typical
prostaglandin,
thromboxane,
and
leukotriene.

The name *prostaglandin* came from the belief that the prostate gland was the compound's source because prostaglandins were first isolated from seminal fluid in 1935. Scientists now know that prostaglandins are produced in a very wide variety of cells. Prostaglandins differ slightly from one another, but they all contain a five-carbon ring. The minor structural differences lead to distinct behaviors, although all prostaglandins lower blood pressure, induce contractions in smooth muscles, and are part of the inflammatory response system.

A number of medications are synthetic prostaglandins. For example, derivatives of the prostaglandin PGE_2 are useful in inducing labor. Prostaglandins associated with inflammation are the main cause of the associated redness, pain, and swelling. The half-life of many prostaglandins is only a few minutes or less. Platelets in the blood generate thromboxanes to serve as vasoconstrictors and to induce aggregation of the platelets, two steps leading to the formation of a blood clot. Thromboxane A_2 is an example of one of these agents that induces blood clotting. White blood cells, leukocytes, and other tissues produce leukotrienes, whose name refers to where they were first discovered (leukocytes) and to the presence of three conjugated double bonds (triene). Leukotrienes are associated with allergy attacks.

Aspirin interferes with the synthesis of prostaglandins and thromboxanes. Aspirin is an anti-inflammatory agent because it counters the inflammation induced by prostaglandins. The interference with the formation of thromboxanes may be part of the reason why low doses of aspirin help prevent heart attacks and strokes. Low thromboxane levels inhibit blood clotting. Another anti-inflammatory drug, cortisone, inhibits the release of arachidonic acid from cell membranes, which, in turn, inhibits the formation of the eicosanoids. The fatty acids in fish oils inhibit the formation of the more potent leukotrienes and thromboxanes.

Chapter 9

Nucleic Acids and the Code of Life

*N*ucleic acids get their name because they were first found in the *nuclei* of cells. DNA *(deoxyribonucleic acid)* — the most famous nucleic acid — is an integral part of *chromosomes,* which contain *genes* that carry the information responsible for the synthesis of proteins. Many of these proteins are *enzymes,* and each catalyzes a specific chemical reaction that occurs in the organism. This ratio gave rise to the *one-gene-one-enzyme* hypothesis, based on the understanding that each gene is responsible for the synthesis of one enzyme. This hypothesis is now better understood as the *one-gene-one-protein* hypothesis because genes give rise to proteins in a wide variety of structural and functional mechanisms within an organism.

DNA has two direct purposes: It must generate new DNA *(replication,* something most people have an interest in) so that new generations of cells have the information necessary to their survival; and it must generate RNA *(ribonucleic acid)* through a process called *transcription.* The RNA is involved in the direct synthesis of proteins, called *translation.* These proteins are essential for the maintenance of life.

Nucleotides: The Guts of DNA and RNA

Both DNA and RNA are polymers of nucleotides. A *nucleotide* is a combination of a nitrogen base, a 5-carbon sugar, and a phosphoric acid. Nucleotides have five different bases and two different sugars. In this section we take a closer look at the components of these nucleotides and show you how they all fit together.

Reservoir of genetic info: Nitrogen bases

The nitrogen bases fall into two categories, the general defining structures of which appear in Figure 9-1.

- ✔ The *purines* (adenine and guanine), composed of two fused rings incorporating two nitrogen atoms in each ring.

- ✔ The *pyrimidines* (cytosine, thymine, and uracil), composed of a single ring with two nitrogen atoms in the ring structure.

Adenine (A), *guanine* (G), and *cytosine* (C) occur in both DNA and RNA. *Thymine* (T) is only found in DNA, whereas *uracil* (U) only occurs in RNA. Modified forms of some of these bases are present in some nucleic acid molecules. The circled hydrogen atoms shown in Figure 9-1 are lost when combining with other components to produce a nucleic acid. The complete structures of the five bases are shown in Figure 9-2. The sequence of these bases stores the genetic information.

The nitrogen and oxygen atoms present on the nitrogen bases provide a number of sites where hydrogen bonding is possible. Hydrogen bonding is most effective and easily formed between certain combinations of nitrogen bases. These combinations are the pattern that's responsible for the transmission of information. The atoms on the nitrogen bases normally use a regular numbering system, whereas the atoms in the sugar component use primed numbers.

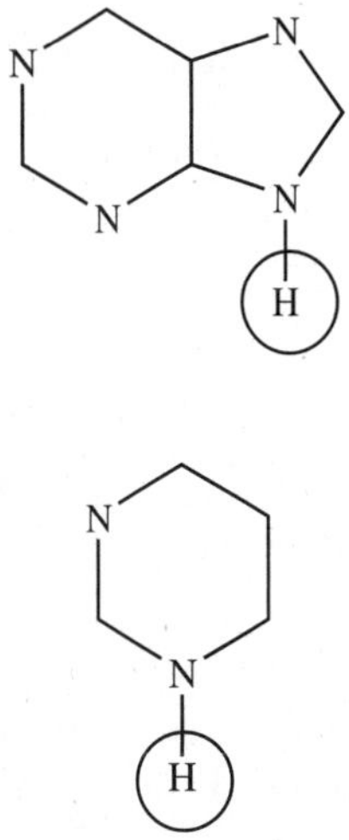

Figure 9-1:
Basic purine structure (top) and basic pyrimidine structure (bottom).

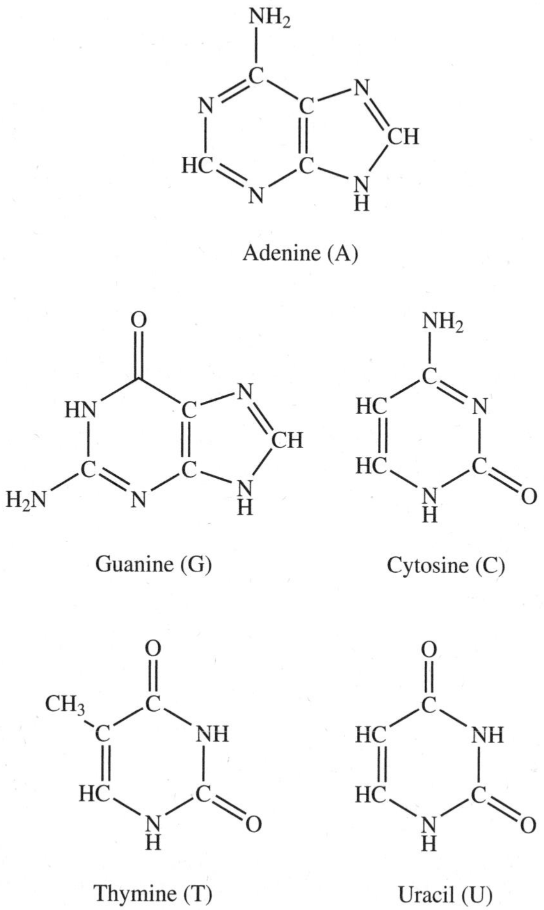

Figure 9-2:
Adenine (A),
guanine (G),
cytosine (C),
thymine (T),
and uracil (U).

Adenine (A)

Guanine (G)   Cytosine (C)

Thymine (T)   Uracil (U)

The sweet side of life: The sugars

The 5-carbon sugars found in the nucleic acids are *D-ribose* and *D-deoxyribose*. The difference between these two sugars is that deoxyribose is missing an oxygen atom on carbon atom number 2'. The structures for these two sugars appear in Figure 9-3. The arrows in the figure point to the alcohol group on carbon atom number 1', the *anomeric* carbon. This is where the nitrogen base attaches. Both sugars adopt the β form of the furanose ring. Numbering of the sugar begins with the anomeric carbon (1') and proceeds clockwise with the $-CH_2OH$ carbon being 5'.

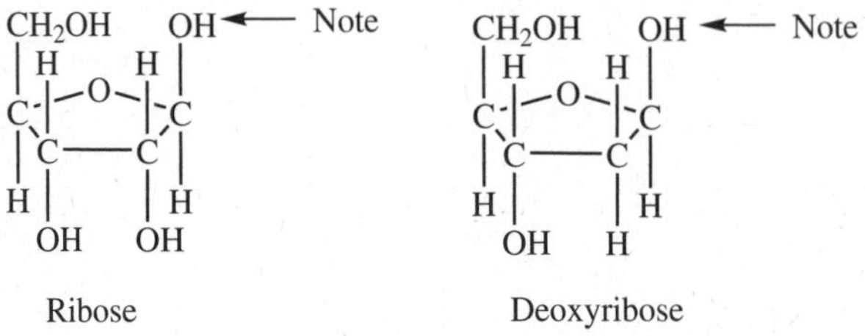

Figure 9-3:
Structures
of the 5-
carbon
sugars
present
in nucleic
acids.

CH₂OH   OH ◄— Note
| H   H |
C-|─O~|-C
| C────C |
H | | H
OH   OH

Ribose

CH₂OH   OH ◄— Note
| H   H |
C-|─O~|-C
| C────C |
H | | H
OH   H

Deoxyribose

The sour side of life: Phosphoric acid

The third component of a nucleotide is a phosphoric acid (see Figure 9-4). At physiological pH it doesn't exist in the fully protonated form shown in the figure. It's responsible for the "acid" in nucleic acid. The different nucleotides link together through their phosphoric acid portions.

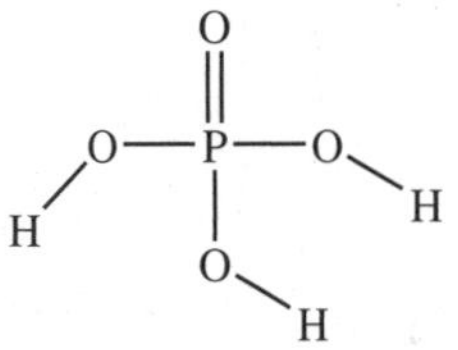

Figure 9-4:
Structure of
phosphoric
acid.

O
‖
O─P─O
H   H
O
H

Tracing the Process: From Nucleoside to Nucleotide to Nucleic Acid

Remember LEGOs and Tinkertoys? Putting together the pieces to get something new? That's what goes on in the construction of nucleic acids. Nature first joins a nitrogen base and a 5-carbon sugar to form a nucleoside; then that nucleoside joins with phosphoric acid to form a nucleotide; finally, the combination of these nucleotides produces a nucleic acid.

First reaction: Nitrogen base + 5-carbon sugar = nucleoside

The combination of a nitrogen base with a 5-carbon sugar is a *nucleoside*. The general reaction appears in Figure 9-5. It's a *condensation* reaction. Remember the condensation reactions you studied in ester formation in organic chemistry? This is exactly the same type. Here a compound containing hydrogen (the nitrogen base) approaches another molecule containing an –OH group (a sugar). The hydrogen combines with the –OH to form water, which is expelled. A bond forms in the remaining fragments.

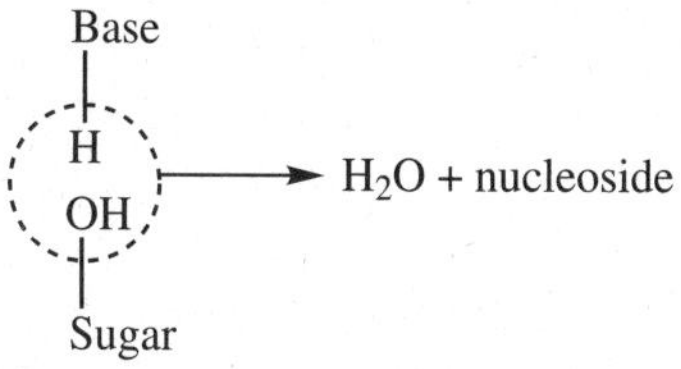

Figure 9-5: General reaction for the formation of a nucleoside.

Base
|
H
OH
|
Sugar

$\longrightarrow$ H_2O + nucleoside

REMEMBER

The nucleoside's name comes from the nitrogen base if the sugar is ribose; it has a prefix if the sugar is deoxyribose. For example, adenine combines with ribose to form *adenosine* and combines with deoxyribose to form *deoxyadenosine.* The structure of the nucleoside adenosine is shown in Figure 9-6.

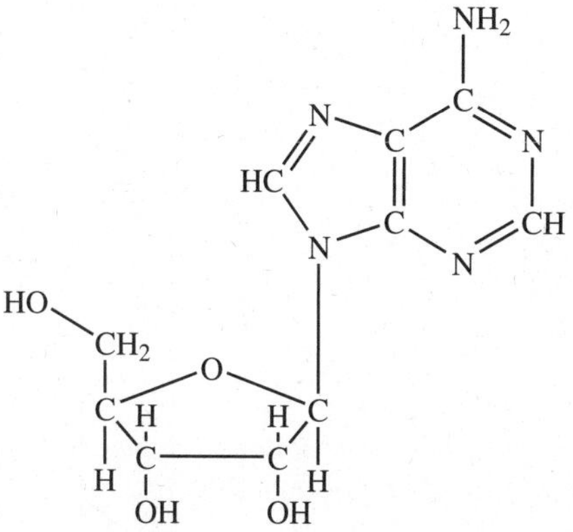

NH$_2$

N   C   N
HC   C   CH
N   C   N

Figure 9-6: Structure of the nucleoside adenosine.

HO
CH$_2$
O
C H   H C
H C   C H
OH   OH

Second reaction: Phosphoric acid + nucleoside = nucleotide

The combination of a phosphoric acid with a nucleoside produces a *nucleotide,* which is a phosphate ester of a nucleoside. The formation involves a condensation reaction between the phosphoric acid and the alcohol group on carbon number 5, the –CH$_2$OH (see Figure 9-7).

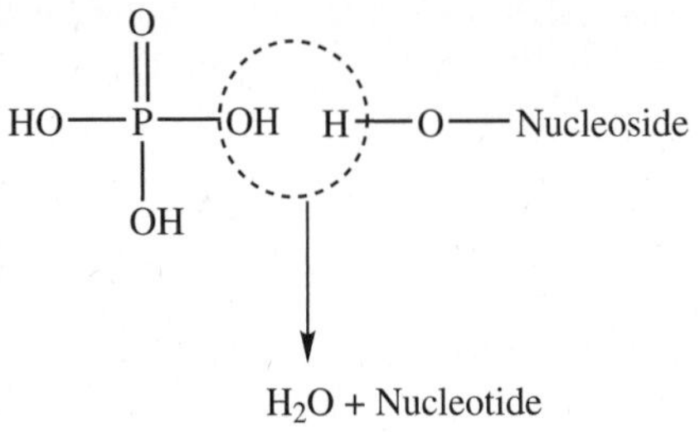

Figure 9-7:
Simplified
representa-
tion of the
formation
of a
nucleotide.

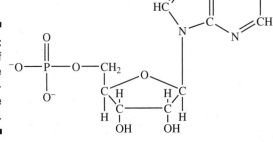

$$H_2O + \text{Nucleotide}$$

Adenosine monophosphate (AMP) is an example of a nucleotide (see Figure 9-8). Nucleotides are the monomers from which nucleic acids form. AMP is one of the building blocks of RNA and is also very much involved in the energy transfer process in the cells (you can read much more about AMP in Part IV).

Figure 9-8:
Structure of
adenosine
monophos-
phate
(AMP).

If the sugar is ribose, the result is one of four *ribonucleotides.* If the sugar is deoxyribose, the result is one of four *deoxyribonucleotides.*

Third reaction: Nucleotide becomes nucleic acid

Nucleic acids form by joining nucleotides using the same condensation reaction we mention in previous sections. This condensation reaction involves the phosphate of one nucleotide reacting with the alcohol group on carbon atom number 3' of another nucleotide (see Figure 9-9). Note that the lower –OH, in the circle, is from the phosphoric acid, attached to carbon-5'. The upper –H in the circle is from the alcohol on carbon-3'.

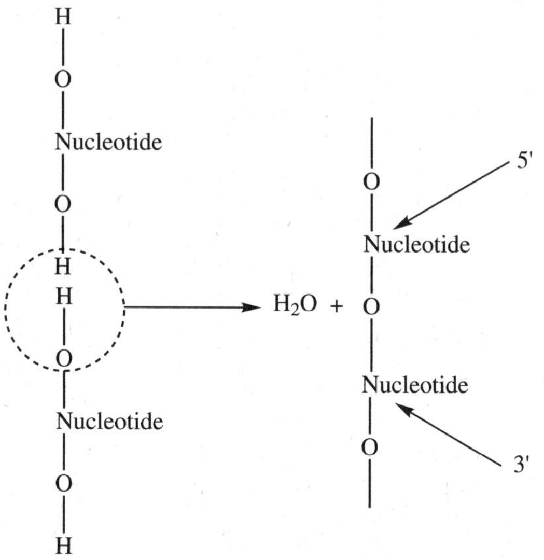

Figure 9-9:
Simplified
representa-
tion of the
joining
of two
nucleotides.

The starting end of the polymer is 5', whereas the terminal end is 3'. Figure 9-10 illustrates the 5' and 3' carbon atoms on adenosine monophosphate.

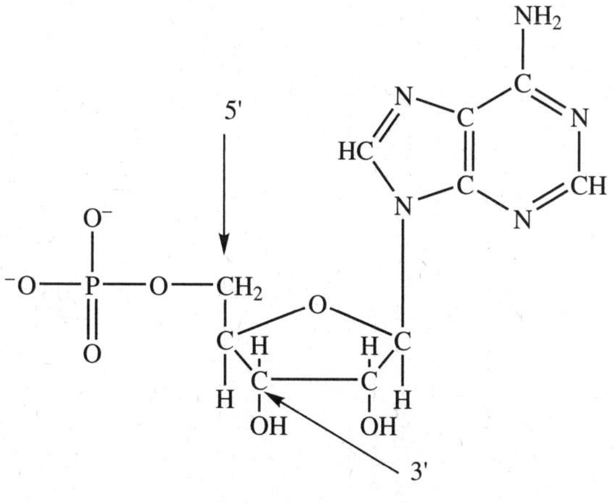

Figure 9-10:
5' and 3'
carbon
atoms on
adenosine
monophos-
phate.

A Primer on Nucleic Acids

Nucleic acids are responsible for storing and directing the information that human cells use for reproduction and growth. They're large molecules found in the cell's nucleus. The genetic information is contained in the DNA, in terms of its primary and secondary structure. As a cell divides and reproduces, the genetic information in the cell is *replicated* to the new cells, which must be done accurately and precisely — no mistakes can be made. RNA's role is to transfer the genetic information found in the DNA to the *ribosomes,* where protein synthesis occurs. DNA and RNA allow humans to live and function. This is known as the *central dogma* in genetics.

DNA and RNA in the grand scheme of life

Both DNA and RNA are polymers composed of nucleotide subunits. However, DNA is a much larger molecule than RNA. DNA molecules typically have molecular weights in the billions. The human genome contains about 3 billion nucleotides.

As a simplification, the structure of a particular nucleic acid may be represented as 5'-C-G-T-A-3'. This abbreviation indicates that you begin at the 5' end and finish at the 3' end (as always), and the nitrogen bases on the nucleotides are, in order, cytosine (C), guanine (G), thymine (T), and adenine (A).

RNA comes in three different types, and each one has a specific use:

- *Ribosomal RNA* (rRNA) is the most common: 75 to 80 percent occurs within the ribosomes of the cell.
- *Transfer RNA* (tRNA) accounts for 10 to 15 percent.
- *Messenger RNA* (mRNA) makes up the remainder.

All three types are important to protein synthesis, which occurs in the ribosomes, home of ribosomal RNA (rRNA). The amino acids necessary for protein synthesis are transferred to the ribosomes by transfer RNA (tRNA). The message instructing the ribosomes how to assemble the protein travels from the DNA to the ribosome via messenger RNA (mRNA). This message tells the ribosome the sequence of amino acids to make a specific protein.

Transfer RNA contains the fewest nucleotides: 70 to 90. The average mRNA has about 1,200 nucleotides. Ribosomal RNA has three subcategories that range from about 120 to over 3,700 nucleotides. (DNA typically has between 1 million and 100 million nucleotides, though viral DNA tends to be smaller.) Ribonucleotides have other uses in addition to building RNA. They're present in energy molecules (ATP), in intracellular hormonal mediators (cyclic AMP), and in certain coenzymes (FAD and NAD+). Plants and animals contain both DNA and RNA. Viruses can contain either DNA or RNA.

Nucleic acid structure

The structure of a particular nucleic acid controls its function within the organism. For example, the structure of a particular tRNA determines which specific amino acid it transfers to the ribosome for protein synthesis. In fact, the difference between DNA and RNA resides in the structure of the molecules. Because of the complexity of these types of molecules, more than one key type of structure may be present.

The primary structure of the nucleic acids is the sequence of nucleotides, the order in which the individual nucleotides are joined. This sequencing determines which hydrogen bonds form, which, in turn, controls much of the nucleic acid's function. DNA also has an important secondary structure, a consequence of hydrogen bonding between the nitrogen bases on the DNA strands. The result is that DNA consists of a *double helix* — which looks like a ladder twisted lengthwise — where hydrogen bonds (the rungs in the ladder) hold the two primary structures together.

The hydrogen bonds between the two strands of DNA make the two strands *complementary (*paired). Every A (adenine) is complementary to a T (thymine), and every G (guanine) is complementary to a C (cytosine) in *base pairing.* Base pairing is essential for the function of the nucleic acids.

The two DNA strands are *antiparallel,* which means that the 5' end of one strand connects to the 3' end of its complementary strand. This pairing also places the more polar (more hydrophilic) sugar and phosphate groups on the outside and the less polar (more hydrophobic) nitrogen bases on the inside. (Note that *hydrophilic* and *hydrophobic* as used here are relative terms.) The antiparallel nature affects how DNA produces new DNA (the replication process) and new RNA (the transcription process).

Although each of the nitrogen bases is very efficient at forming hydrogen bonds, certain combinations are extremely effective. In DNA, an adenine is capable of forming two hydrogen bonds to thymine (see Figure 9-11), and guanine can form three hydrogen bonds to cytosine (see Figure 9-12).

Adenine is also able to form hydrogen bonds with uracil when DNA interacts with RNA or when two RNA molecules interact. The interaction between adenine and uracil is shown in Figure 9-13.

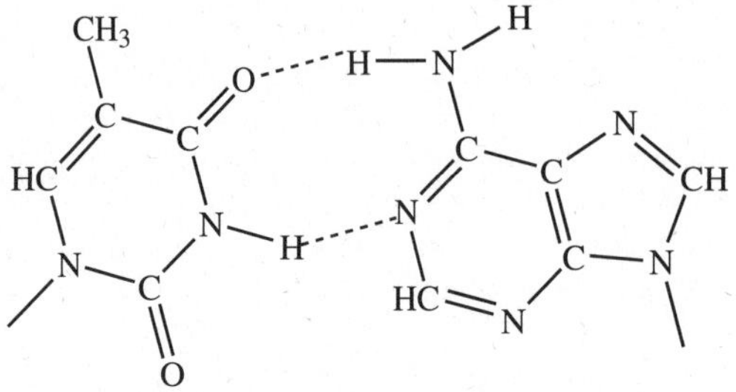

Figure 9-11: Hydrogen bonds (dotted lines) form between adenine (right) and thymine (left).

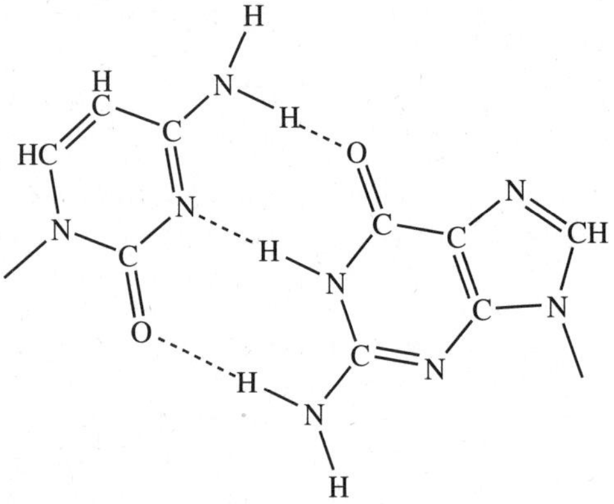

Figure 9-12: Hydrogen bonds (dotted lines) form between guanine (right) and cytosine (left).

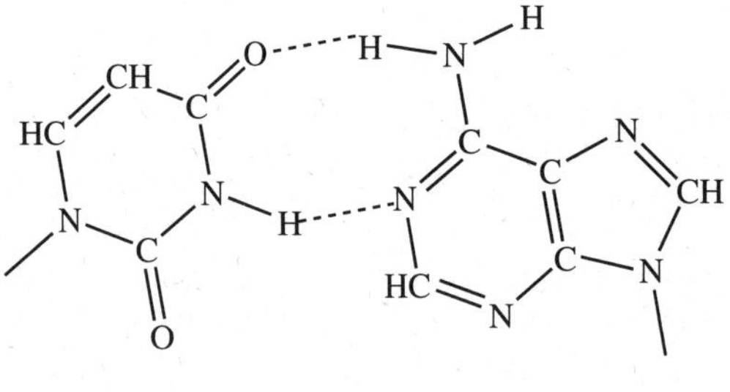

Figure 9-13: Hydrogen bonds (dotted lines) form between adenine (right) and uracil (left).

The ability to form these specific combinations is important in real life; this is the *genetic code* you've heard so much about. The sequencing of nucleotides in the nucleic acids and the sequencing of amino acids in the proteins all depend on these hydrogen bonds. Without them, the appropriate information wouldn't be transferred precisely, and you may produce kittens instead of kids. How does nature pass on that genetic code? Through DNA, the structure of life (see Figure 9-14).

Figure 9-14:
The secondary structure of DNA.

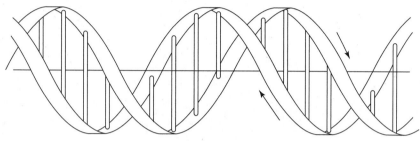

Chapter 10

Vitamins: Both Simple and Complex

In This Chapter

- Taking a look at the purpose of vitamins
- Understanding B vitamins
- Assessing other vitamins and nutrients

An organism must absorb a variety of materials to live, many of which fall into the category of food, certainly one of *our* favorite categories, especially John's. These foodstuffs required by an organism for life and growth are classified as nutrients. *Nutrients* are the substances in the diet necessary for growth, replacement, and energy. Here are the six general classes of nutrients:

- Carbohydrates
- Lipids
- Minerals
- Proteins
- Vitamins
- Water

Digestion converts large molecules in food into smaller molecules that the body can absorb. During digestion, the body breaks down carbohydrates (with the exception of the monosaccharides), lipids, and proteins into their components (we cover these nutrients earlier in this book). The organism often uses these components directly for growth and replacement. Animals get their energy primarily from carbohydrates and lipids, but proteins can also serve as an energy source.

Organisms also need the organic materials of vitamins and the inorganic materials of minerals. In addition, all living organisms require water to survive. Water is a wonderful substance. For more about the unusual properties of water, check out Chapter 2 in this book or *Chemistry For Dummies* by John T. Moore (Wiley).

More than One-a-Day: Basics of Vitamins

Vitamins are organic compounds that are required, in small quantities, for normal metabolism. The term *active form* is used to describe the structural form of the molecule, in this case vitamins, that performs its function (exhibits *activity*) within the organism. In general, humans can't synthesize sufficient quantities of vitamins; thus, vitamins must come from other sources — through the diet and/or in pill form. A deficiency of a vitamin in the diet leads to a health problem. The general symptoms for any vitamin deficiency include frequent illness, slow healing of wounds, and tiredness. It wasn't until the early 1900s that the need for trace nutrients such as vitamins and minerals was first established.

Vitamins fall into two categories: water-soluble and fat-soluble. Water-soluble vitamins tend to have more oxygen and nitrogen in their structure than fat-soluble vitamins, which have significant hydrocarbon portions in their structure. *Water-soluble* vitamins include vitamin C and the B vitamins. Vitamins A, D, E, and K comprise the other category, the *fat-soluble* vitamins. The majority of water-soluble vitamins either act as coenzymes or are important in the synthesis of coenzymes. Fat-soluble vitamins serve a variety of biochemical functions.

The body can easily eliminate an excess of the water-soluble vitamins, normally in the urine. The bright yellow of the urine of a person taking large doses of vitamin C attests to that fact. Because the body doesn't store water-soluble vitamins, continual replacement is necessary. The body can store excess amounts of a fat-soluble vitamin in the body's fatty tissue, and therefore, elimination isn't very easy. Unfortunately, this can lead to an accumulation of these vitamins, sometimes to toxic levels. One should consider this before consuming mega quantities of the fat-soluble vitamins.

To B or Not to B: B Complex Vitamins

The *B vitamins* — or *B complex* — comprise a number of water-soluble vitamins that are found together in a number of sources. Originally, this mixture was thought to be only one vitamin (vitamin B). With the possible exception of vitamin B_6, these appear to be relatively nontoxic. In general, the B complex is important for healthy skin and nervous systems.

Vitamin B_1 (thiamine)

Thiamine (vitamin B_1) is important to carbohydrate metabolism. Like the other B vitamins, the body doesn't store it. In addition, prolonged cooking of food can destroy it. After the body absorbs thiamine, it converts thiamine to a form that's biologically active through the attachment of a pyrophosphate (diphosphate) group to give thiamine *pyrophosphate* (TPP). Figure 10-1 shows the structures of vitamin B_1 and thiamine pyrophosphate.

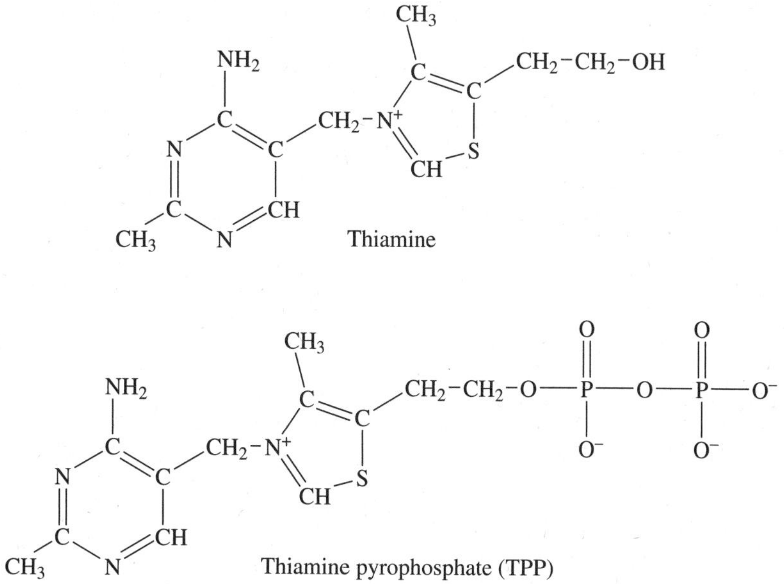

Figure 10-1:
Structures of vitamin B1 (thiamine) and thiamine pyrophosphate (TPP).

Thiamine

Thiamine pyrophosphate (TPP)

TPP is a coenzyme used in decarboxylating pyruvate to acetyl-CoA and A-ketoglutarate to succinyl-CoA. In addition, TPP is necessary for the synthesis of ribose.

A deficiency in thiamine leads to *beriberi,* which causes deterioration in the nervous system. Beriberi was prevalent in regions where rice was a major food source. Rice, particularly polished rice, is low in thiamine. Using brown rice, which has more thiamine, alleviates this problem. Nursing infants are particularly at risk when their mothers have a thiamine deficiency. Many alcoholics also suffer from this condition because many "foods" high in alcohol are particularly low in vitamins.

Good dietary sources of thiamine include liver, spinach, green peas, navy and pinto beans, whole-grain cereals, and most legumes.

Vitamin B₂ (riboflavin)

Riboflavin (vitamin B_2) is essential to the synthesis of flavin mononucleotide (FMN) and flavin adenine dinucleotide (FAD). Figure 10-2 shows the structures of these materials. FMN and FAD are important coenzymes involved in a number of biochemical redox processes. The name *riboflavin* alludes to the presence of *ribitol,* an alcohol derived from ribose. The other part of riboflavin is the ring system isoalloxazine, a flavin derivative.

No deficiency diseases are associated with riboflavin; however, a deficiency does lead to burning and itchy eyes, dermatitis, and anemia. Dietary sources of this vitamin include soybeans, liver, milk, cheese, and green leafy vegetables. Riboflavin is stable during cooking but is broken down by light.

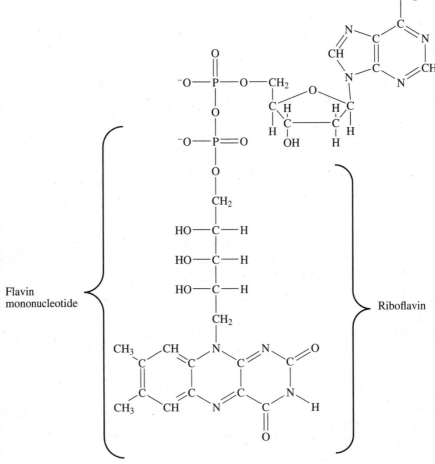

Figure 10-2:
Structure of flavin adenine dinucleotide (the entire structure) and the component materials flavin mononucleotide and riboflavin.

Vitamin B₃ (niacin)

The term *niacin* (vitamin B₃) applies to two compounds: nicotinic acid and nicotinamide. These two compounds, along with nicotinamide adenine dinucleotide (NAD⁺), appear in Figure 10-3. Nicotinamide is part of the coenzymes NAD⁺ and nicotinamide dinucleotide phosphate (NADP⁺). These coenzymes work with a number of enzymes in catalyzing a number of redox processes in the body.

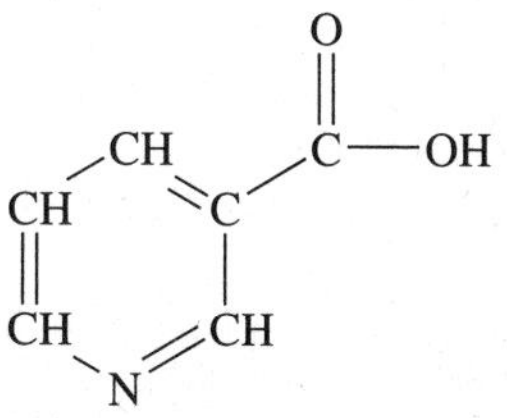

Figure 10-3: Structures of nicotinic acid, nicotinamide, and nicotinamide adenine dinucleotide (NAD⁺).

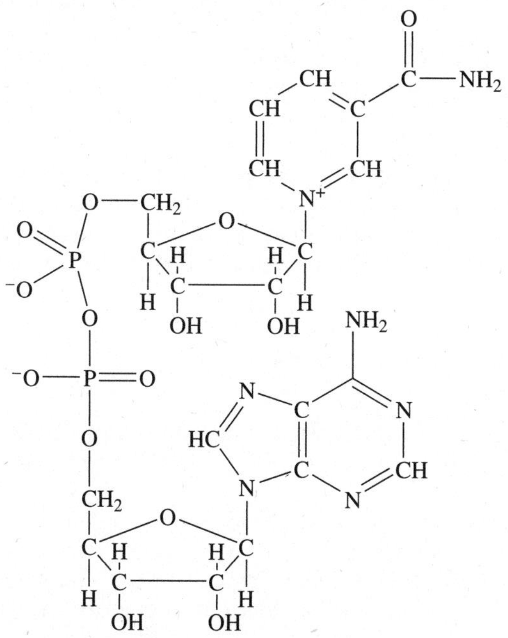

Nicotinic acid

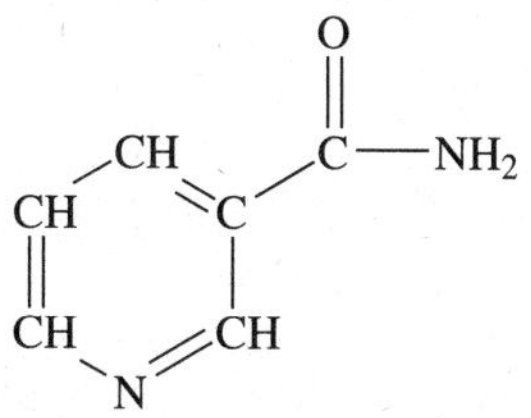

Nicotinamide

Nicotinamide adenine dinucleotide (NAD⁺)

Niacin is one of the few vitamins that the body *can* synthesize. The synthesis utilizes tryptophan and isn't very efficient.

Pellagra is a niacin-deficiency disease. Symptoms include loss of appetite, dermatitis, mental disorders, diarrhea, and possibly death. It was common in the southern United States in the early 1900s because many people had a diet of corn, which isn't a good source of niacin or tryptophan.

You can find many dietary sources for niacin, including most meats and vegetables, milk, cheese, and grains.

Vitamin B₆ (pyridoxine)

Vitamin B₆ consists of three components: pyridoxine, pyridoxal, and pyridoxamine. All three need to be converted to pyridoxal phosphate, a form that's biologically active in the organism. Figure 10-4 shows the structures for these compounds. Pyridoxal phosphate serves as a coenzyme in a variety of processes, including the interconversion of A-keto acids and amino acids.

Avocados, chicken, fish, nuts, liver, and bananas are especially good food sources of vitamin B_6. Heating decreases its concentration in food. (But don't try eating raw chicken just for the increased B_6.)

No pyridoxine-deficiency diseases are known, but low levels can lead to irritability, depression, and confusion. Unlike the other water-soluble vitamins, evidence shows that large doses of vitamin B_6 may lead to health problems. The symptoms of excess vitamin B_6 consumption include irreversible nerve damage.

Biotin

Biotin (see Figure 10-5) is a coenzyme important to many carboxylation reactions. Biotin is the carbon transporter in both lipid and carbohydrate metabolism.

Bacteria in the intestinal track synthesize biotin in sufficient quantities to minimize the chances for a deficiency. However, antibiotics can inhibit the growth of these bacteria and induce a deficiency. In these circumstances, the symptoms include nausea, dermatitis, depression, and anorexia. Biotin is stable to cooking.

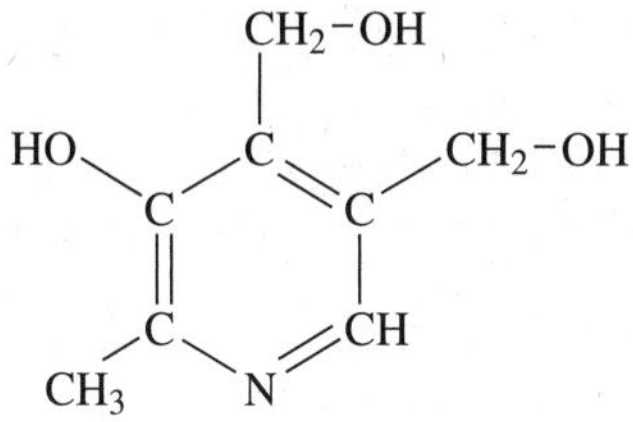

Figure 10-4:
Structures
of
pyridoxine,
pyridoxal,
pyridoxamine,
and
pyridoxal
phosphate.

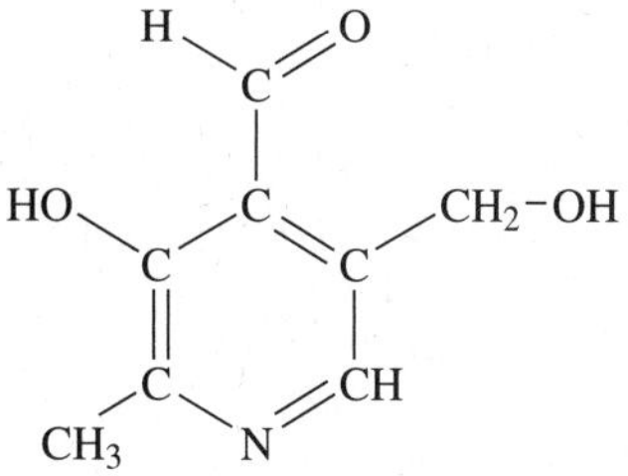

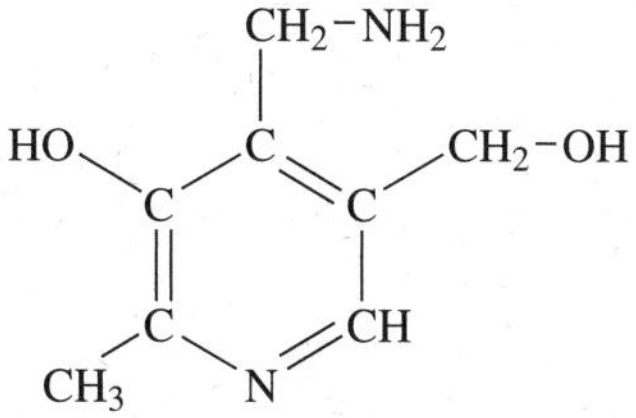

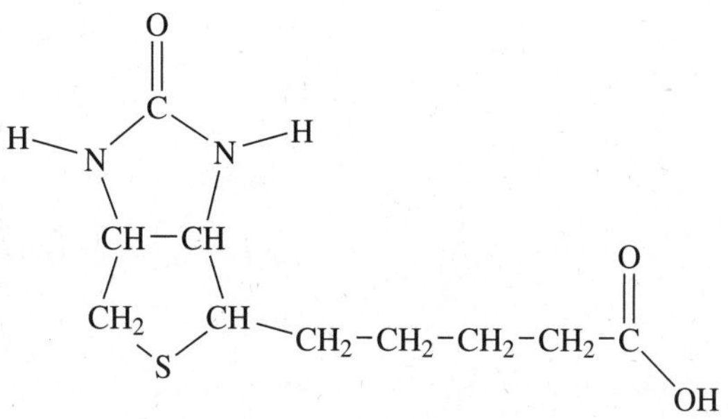

Pyridoxine

Pyridoxal

Pyridoxamine

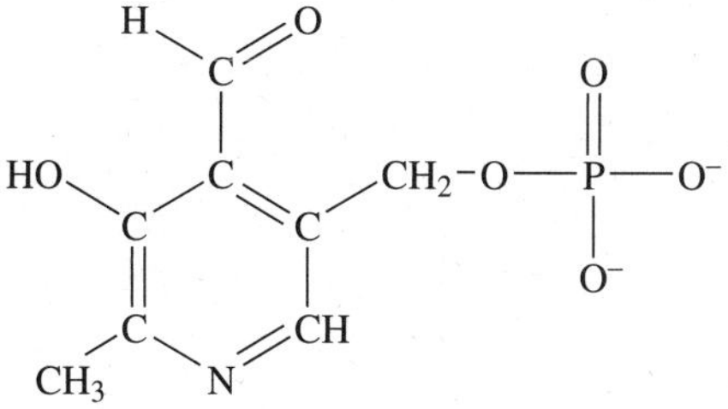

Pyridoxal phosphate

Figure 10-5:
Structure of
biotin.

Folic acid

Bacteria in the intestinal track also produce *folic acid;* however, green leafy vegetables, dried beans, and liver (one of John's favorites) are also sources. Reduction of folic acid yields tetrahydrofolic acid, the active form. Figure 10-6 shows both structures. The coenzyme transports a carbon, usually as a methyl or formyl, in the synthesis of heme, nucleic acids, choline, and several other compounds. Although cooking easily destroys the compound, intestinal bacteria normally produce sufficient quantities.

Folic acid is critical to the prevention of malformations of the brain *(anencephaly)* and spine *(spina bifida)*. A deficiency of folic acid affects the synthesis of purines; symptoms include gastrointestinal disturbances and anemia. Pregnant women are normally advised to take a vitamin high in folic acid to help in the normal development of the fetus, especially the spine and brain. Sulfa drugs interfere with the formation of folic acid by some pathogens via a form of competitive inhibition.

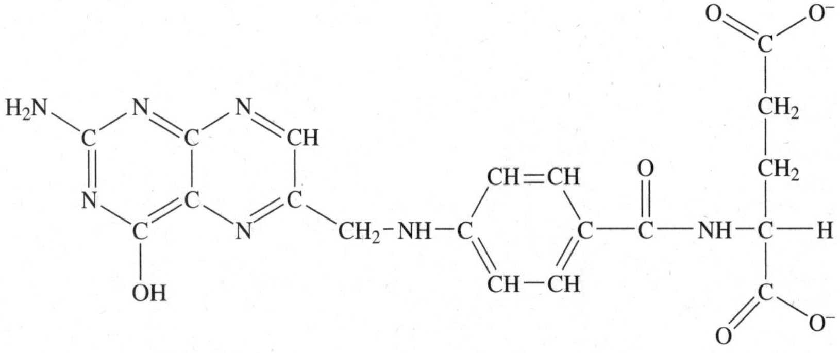

Folic acid

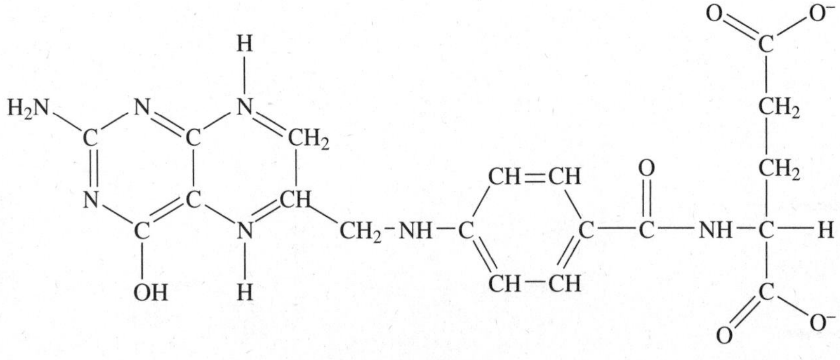

Figure 10-6:
Structures
of folic
acid and
tetrahydro-
folic acid.

Tetrahydrofolic acid

Pantothenic acid

The name *pantothenic acid* (see Figure 10-7) derives from a Greek word meaning "from everywhere." As you may expect, then, it has numerous sources, including whole grains, eggs, and meat. Deficiency is virtually unknown. The vitamin isn't destroyed by moderate cooking temperatures, but it also isn't stable at high cooking temperatures.

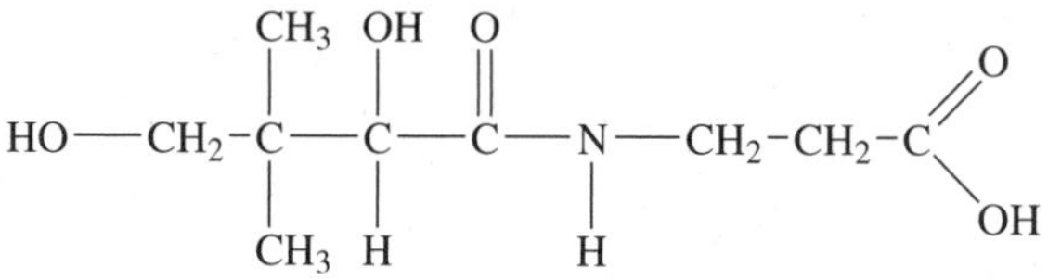

Figure 10-7:
Structure of
pantothenic
acid.

 Pantothenic acid is necessary in the biosynthesis of coenzyme A. Coenzyme A is an exceedingly important substance in many biological processes because it transfers acyl groups.

The wonders of vitamin B_{12}

Vitamin B_{12} is the only known natural organometallic compound. It doesn't occur in higher plants, and apparently only bacteria are capable of synthesizing it — bacteria that live in their hosts in a symbiotic relationship. Unfortunately, higher animals including human beings don't have these types of bacteria. Thus, humans must obtain vitamin B_{12} from food.

The name *cyanocobalamine* refers to the presence of cyanide. The cyanide is an artifact of the isolation of the compound and isn't naturally present. Vitamin B_{12} is necessary to the formation of two coenzymes: methylcobalamin (see Figure 10-8) and 5'-deoxyadenosylcobalamin.

Both coenzymes assist in reactions involving rearrangements. Methylcobalamin is useful in methyl transfer reactions. The coenzyme 5'-deoxyadenosylcobalamin works in some rearrangement reactions where a hydrogen atom and a group attached to an adjacent carbon exchange positions.

Pernicious anemia usually results from poor absorption of vitamin B_{12}. Normal stomach cells produce a glycoprotein that aids in the absorption of the vitamin in the intestine. The lack of this intrinsic factor, not the lack of the vitamin in the diet, leads to the vitamin deficiency. Elderly people may have difficulty in generating sufficient quantities of the intrinsic factor, and strict vegetarians also may develop symptoms. The symptoms of pernicious anemia include lesions on the spinal cord leading to a loss of muscular coordination and gastrointestinal problems. The blood contains large, fragile, and immature red blood cells. Dietary sources include meat, eggs, milk, and cereals. This vitamin is stable to cooking.

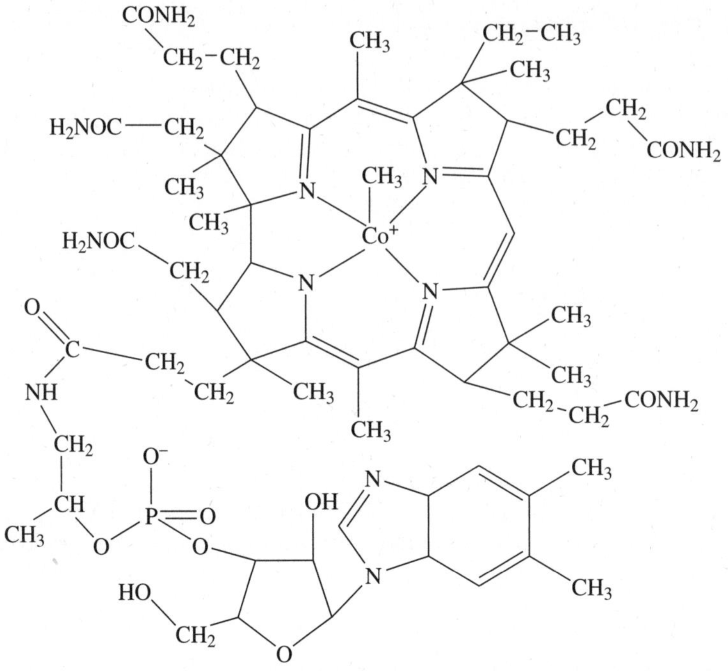

Figure 10-8: Structure of methyl-cobalamin.

Vitamin A

Vitamin A isn't a single compound. A number of compounds are *biologically active* — that is, they undergo biological reactions within the organism. The parent compound is 11-trans-retinol, found in milk and eggs. Vitamin A is exclusive to animals, and the plant pigment β-carotene can serve as a precursor (see Figure 10-9). As a precursor, it's a *provitamin*. Cleavage of β-carotene yields two vitamin A active species. Any β-carotene that doesn't become vitamin A is used as an antioxidant.

Vitamin A is especially important to vision. Part of the vision process involves the absorption of light. This absorption causes the geometry on the double bond between carbon atoms 11 and 12 to change from cis to trans. The isomerization triggers a series of events, giving rise to a nerve impulse. An enzyme reverses the isomerization so the molecule may be reused. In addition to being directly involved in vision, vitamin A also promotes the development of the epithelial cells that produce the mucous membranes, which protect the eyes and many other organs from infections and irritants. Vitamin A also helps in the changes in the bone structures that occur as an infant matures.

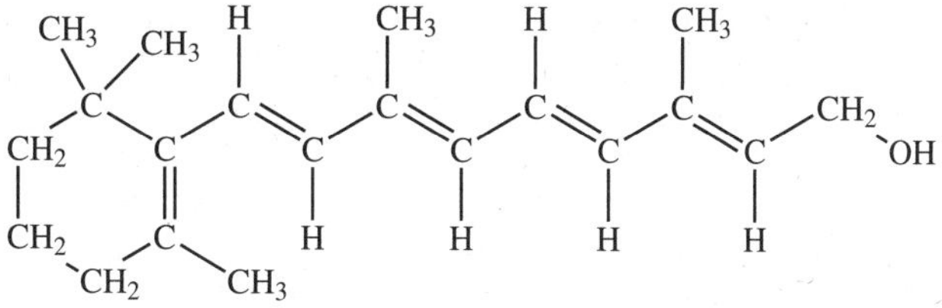

Figure 10-9:
Structures
of 11-trans-
retinol and
β-carotene.
Note that
carbon 11
is the fifth
from the
right in the
main chain.

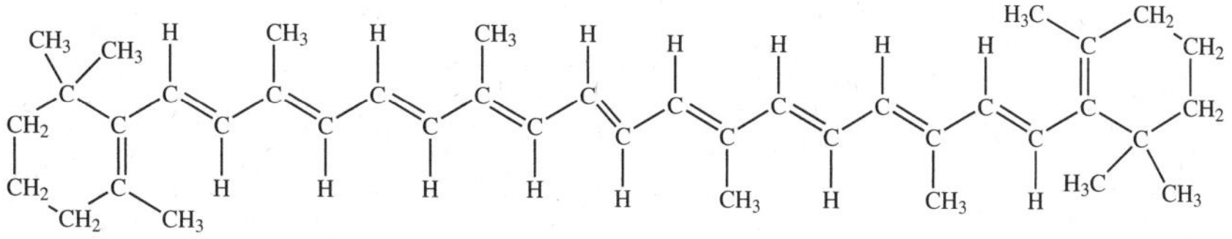

11-trans-retinol

β-carotene

A deficiency in vitamin A begins with night blindness, followed by other eye problems, which can lead to blindness. An extreme deficiency may lead to *xerophthalmia,* inflammation of the eyelids and eyes, which can cause infections and blindness. Young animals require vitamin A for growth, and adults are capable of storing several months' supply of it, primarily in the liver. The livers of some animals, such as polar bears and seals, may have such a high vitamin A concentration that they're toxic to humans. (We weren't planning on munching on a polar bear liver sandwich anyway.) Excessive dosages of vitamin A may lead to acute toxicity, and as a fat-soluble vitamin, it isn't easily eliminated. Symptoms include nausea, vomiting, blurred vision, and headaches. Large doses have been linked to birth defects and spontaneous abortions. The provitamin, β-carotene, isn't toxic, although it is fat-soluble. It's found in carrots, and eating a *lot* of carrots can lead to an orange/yellow skin color. (We didn't notice that ever happening to Bugs Bunny.)

Vitamin C

Vitamin C is another name for *ascorbic acid* (see Figure 10-10). Dehydroascorbic acid also has vitamin C activity. Vitamin C is water-soluble — thus the body can readily eliminate excess, and large doses are not toxic. Vitamin C is an antioxidant. Like vitamin E, it helps prevent damage produced by oxidants. It also helps in the absorption of iron and keeps the iron in the +2 state. Vitamin C helps convert some of the proline in collagen C to hydroxyproline, which stabilizes the collagen.

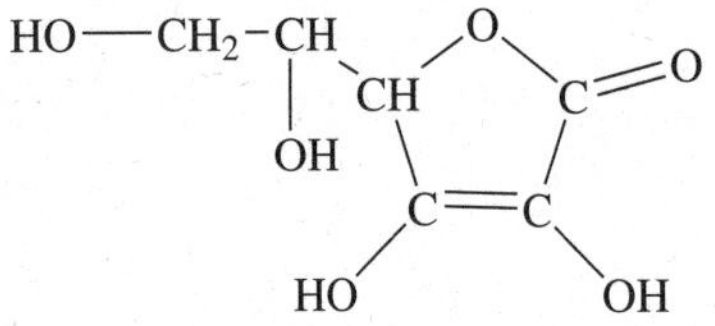

Figure 10-10:
Structure of
vitamin C.

$$HO-CH_2-CH \quad \overset{\displaystyle O}{\underset{CH}{\overset{|}{\underset{OH}{}}}} \quad C=O$$

A deficiency in vitamin C leads to the disease *scurvy,* symptoms of which include a weakening of the collagen, an important protein in connective tissues such as ligaments and tendons. Many foods contain vitamin C, especially plants and citrus fruits, so preventing scurvy is easy. For years, British ships carried limes as a source of vitamin C (leading, incidentally, to the slang term *limey* to refer to a British sailor). Many mammals (other than humans and other primates) synthesize vitamin C from glucose. Cooking, especially prolonged cooking, destroys vitamin C.

Vitamin D

Vitamin D is sometimes called the *sunshine vitamin.* The body can produce it through the action of sunlight, which is ultraviolet radiation. Individuals who walk around outside nude or semi-nude (not a pretty picture for many of us) normally have very little trouble with vitamin D deficiency. Actually, 15 minutes a day outside in shorts and a T-shirt is enough to take care of your vitamin D needs. Still, most people still depend on vitamin D-fortified foods, especially milk.

Several compounds exhibit vitamin D activity. Only two of them — actually provitamins — occur commonly in food: ergosterol and 7-dehydrocholesterol. Irradiation with ultraviolet light converts ergosterol into vitamin D_2, *ergocaliferol.* Ultraviolet irradiation, particularly in the skin of animals, converts 7-dehydrocholesterol into vitamin D_3, *cholecalciferol.* (A little confusingly, vitamin D_1 is a mixture of vitamin D_2 and vitamin D_3.) The structures of ergosterol and vitamin D_2 are shown in Figure 10-11; 7-dehydrocholesterol and vitamin D_3 appear in Figure 10-12.

The body's ability to absorb calcium and phosphorus is tied to vitamin D. Teeth and bone have large amounts of these two elements and are the first parts of the body affected by a vitamin D deficiency. *Osteomalacia,* a condition in which a softening of the bones may lead to deformities, may also result. (In infants and children, osteomalacia is called *rickets.*) A vitamin D deficiency is more serious in children than in adults because growth requires larger quantities of calcium and phosphorus. Persons with some portion of their skin routinely exposed to sunlight seldom develop a deficiency. However, with the increased use of high SPF sunblocks, physicians are actually seeing deficiencies in vitamin D more often.

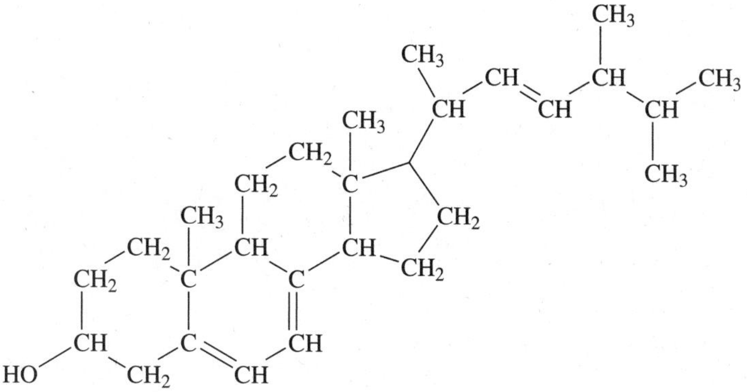

Ergosterol

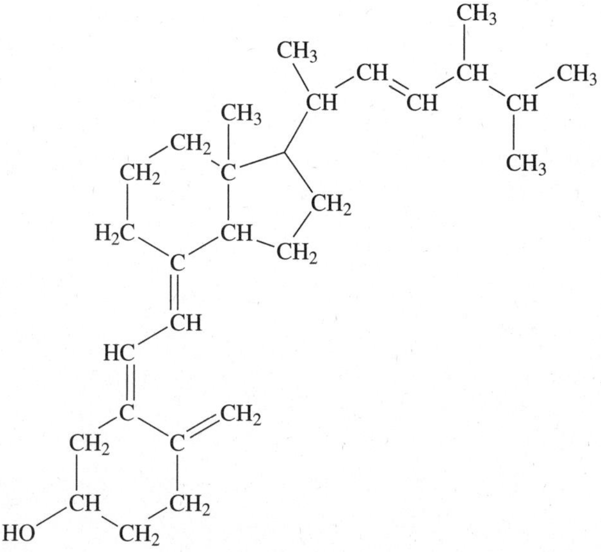

Figure 10-11:
Structures of ergosterol and vitamin D$_2$.

Vitamin D$_2$

Excess vitamin D is toxic. Eliminating this fat-soluble vitamin from the body isn't easy. Symptoms of excessive amounts of vitamin D include nausea, diarrhea, kidney stones and other deposits, and sometimes even death.

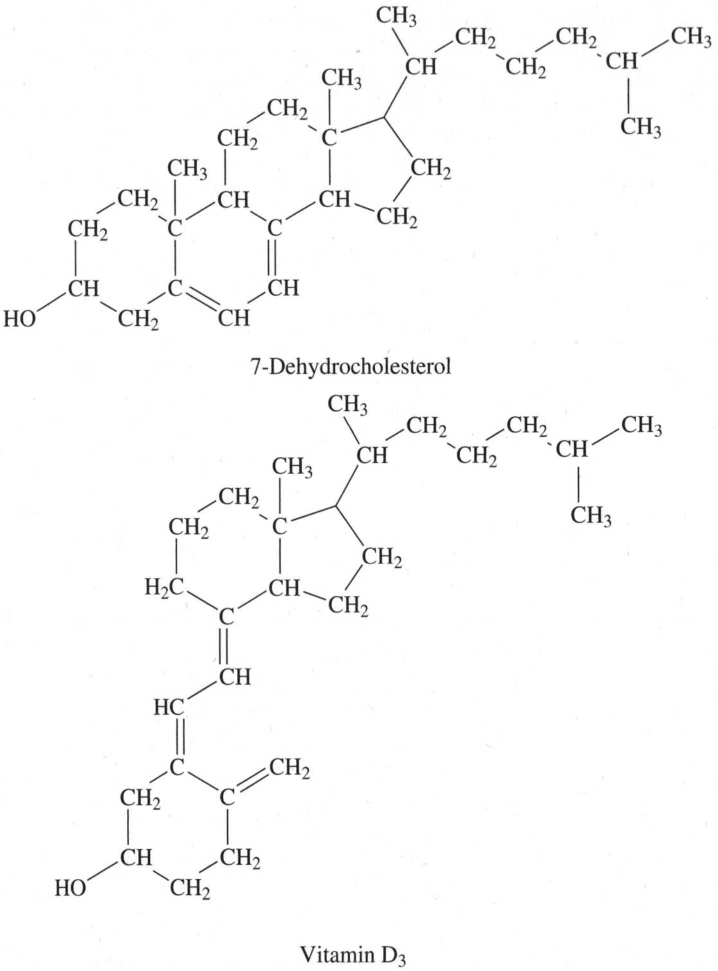

7-Dehydrocholesterol

Figure 10-12:
Structures
of 7-
dehydro-
cholesterol
and
vitamin D₃.

Vitamin D₃

Vitamin E

The *tocopherols* are a group of compounds that exhibit vitamin E activity. The most effective is α-tocopherol (see Figure 10-13). Vitamin E comes from a number of sources — vegetable oils, nuts, whole grains, and leafy vegetables, to name a few. Deficiencies are rare except in individuals on a no-fat diet or those who, for medical reasons, can't efficiently absorb fat. Cystic fibrosis may interfere with fat absorption.

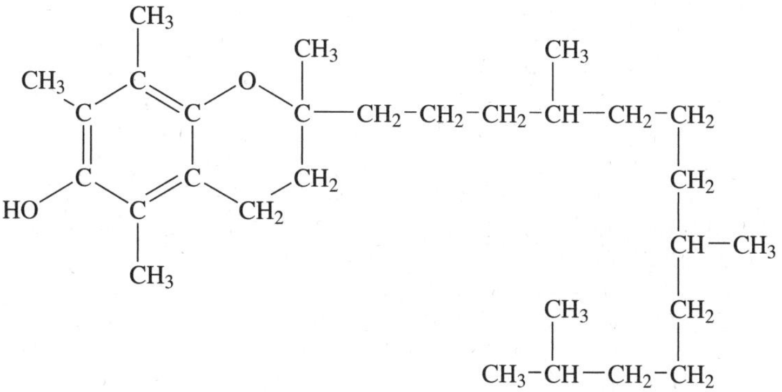

Figure 10-13:
Structure
of α-
tocopherol
(vitamin E).

Vitamin E serves as an effective antioxidant. *Antioxidants* are necessary to minimize the damage caused by oxidants present in the body — many problems associated with aging are apparently due to oxidants. Vitamin E may also help prevent cholesterol deposits in the arteries. No problems with the use of large doses of vitamin E have been documented; however, some recent studies, although disputed, warn against taking mega doses of vitamin E.

Vitamin K

Vitamin K₁ (see Figure 10-14) is one of many compounds that exhibit vitamin K activity. Vitamin K variants differ in the side chains attached to the ring system. One chain is usually a methyl; the other typically has at least 20 carbon atoms.

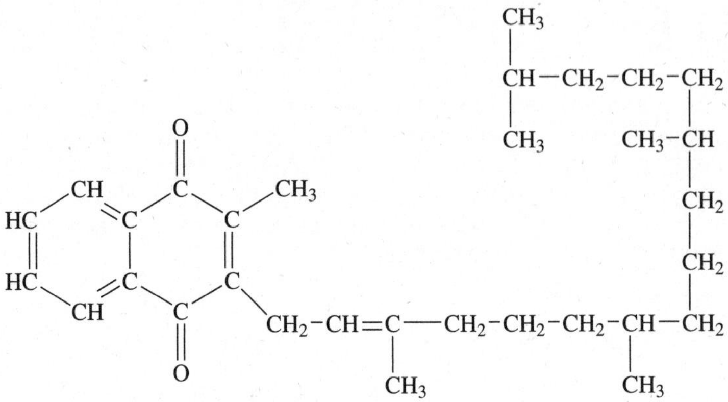

Figure 10-14:
Structure of
vitamin K$_1$.

Vitamin K is necessary to produce the proenzyme *prothrombin,* which helps blood clot. A vitamin K deficiency is uncommon because intestinal bacteria normally produce sufficient quantities, although several foods are also good sources, including green leafy vegetables, cauliflower, broccoli, organ meats (love that liver!), milk, soybeans, avocados, and bananas. Two tablespoons of parsley contain almost twice your recommended daily amount of vitamin K. (You can eat parsley? I thought it was just for decoration!) Prolonged use of antibiotics can decrease the number of these vitamin K-producing bacteria and lead to a reduction in vitamin K in the body. One symptom of a deficiency is an increase in the time necessary to form a blood clot, and such individuals are prone to develop serious bruises from even minor injuries. Infants with a deficiency have been known to die from brain hemorrhage. Increasing the vitamin K intake of the mother decreases the likelihood of this occurrence.

Vitamin K also has a use in the pet world. If a pet accidentally eats rat poison (which prevents clotting), a vet is likely to give the pet a really large dose of vitamin K to counteract the effects of the rat poison.

Chapter 11

Hormones: The Body's Messengers

. . . .

In This Chapter

▶ Examining the structures of hormones

▶ Activating certain prohormones

▶ Discovering how hormones function

. . . .

*H*ormones are substances produced in one area of the body and used in another. They're molecular messengers that are created in specific glands in the body, which then travel through the bloodstream to the hormone's target organ. Some hormones may exhibit *autocrine* or *paracrine* function, such as *growth and differentiation factors* (GDFs), which convey biochemical information within a particular organ (a bit like passing a note in class). This conveyance is accomplished by simple diffusion over a small distance. Some of these molecular messengers may travel via the bloodstream (endocrine) or short distances among cells (paracrine, autocrine).

The endocrine glands produce most — but not all — hormones. Endocrine glands include (but are not limited to) the hypothalamus, pituitary, pancreas, adrenal, liver, testes, and ovaries. Now surely that got your attention! Some glands produce a single hormone, whereas others produce more than one. The simplified viewpoint (and we are all about keeping it simple) is that the pituitary gland acts as the central control for the endocrine system. Hormones from the pituitary gland do cause other glands to produce hormones that affect other systems. However, some glands have the same effect on the pituitary gland.

Structures of Some Key Hormones

The three groups of hormones are

- **Proteins,** such as insulin
- **Steroids,** materials derived from cholesterol
- **Amines,** such as epinephrine

These materials allow one part of the body to influence what occurs elsewhere. These molecules are so efficient that only very low concentrations, typically 10^{-7} to 10^{-10} M, are necessary to influence what is happening elsewhere in the body. That's a really small amount! The low concentrations make identification and isolation of these substances difficult.

In general, protein and amine hormones are hydrophilic, while the steroid hormones tend to be lipophilic. Recall that the lipid bilayer, which constitutes the cell membrane, is hydrophobic; this inhibits hydrophilic hormones from simply entering the cell. In many cases, stimulation due to these hormones is the result of the hormone's interaction with receptor sites on the cell's exterior. The interaction with the receptor site induces the generation of a second messenger. Lipophilic hormones usually transverse the cell membrane via passive diffusion. Once through the cell membrane, these hormones can travel to the receptor site.

Proteins

The *protein,* or *polypeptide,* hormones, typically produced by the pituitary and hypothalamus glands, vary greatly in size — from simple tripeptides to larger molecules with more than 200 amino acid residues. Protein hormones are a diverse collection of molecules, including insulin (the structure of which you can see in Chapter 5).

One important protein hormone is the *thyrotropin-releasing factor,* which induces the release or production of a biochemical (thyrotropin, in this case). The thyrotropin-releasing factor hormone is a tripeptide that contains glutamine (modified), histidine, and proline (modified). Another important protein hormone is the *growth-hormone-release-inhibitory factor,* which inhibits the release or production of a chemical species. Together (see Figure 11-1), these types of hormones provide a mechanism to start and stop an action. The idea is to maintain a tight biochemical control of biochemical processes, such as growth.

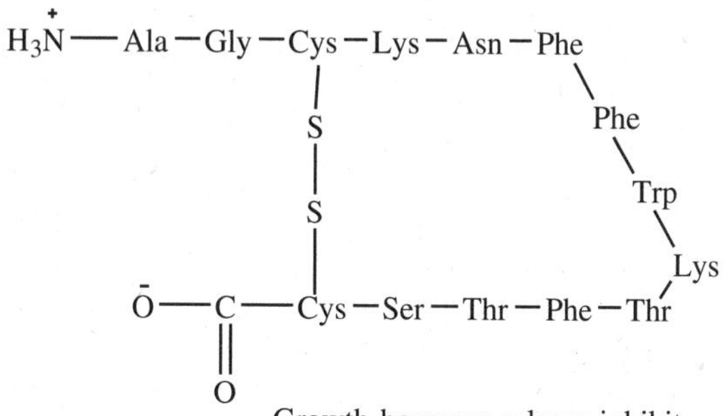

Growth-hormone-release-inhibitory factor

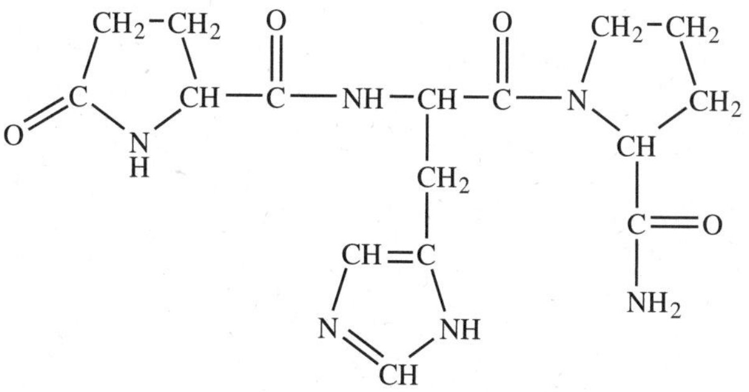

Figure 11-1:
Structures
of the
growth-
hormone-
release-
inhibitory
factor
and the
thyrotropin-
releasing
factor.

Thyrotropin-releasing factor

Steroids

You've no doubt read about steroid use among athletes, who use these substances to increase muscle mass — to "pump-up," in other words. *Steroid* hormones, produced by the body's ovaries, testes, and adrenal glands, are cholesterol derivatives of about the same size as the parent molecule. They include the *estrogens* (female sex hormones), the *androgens* (male sex hormones), and the *adrenal cortical* hormones, such as *aldosterone* and *cortisol.* The estrogens and androgens are responsible for the development of the secondary sex characteristics of females and males, respectively. These characteristics include enlargement of the breasts in females and development of facial hair in males.

The adrenal cortical hormones (see Figure 11-2), which include the *glucocorticoids* and the *mineralocorticoids,* have a variety of functions. The glucocorticoids, such as cortisol, are important to several metabolic pathways and are often used in the stress response (we bet you'll have a cortisol spike during your biochemistry exam!). The mineralocorticoids, such as aldosterone, are important to the transport of inorganic species, such as sodium or potassium ions.

Amines

The *amine* hormones, typically produced by the thyroid and adrenal glands, are small molecules, many of which are derivatives of tyrosine. These hormones include *thyroxine* and *triiodothyronine,* produced by the thyroid gland, and *epinephrine* and *norepinephrine,* produced by the adrenal gland. Figure 11-3 illustrates the structures of these hormones. Thyroxine and triiodothyronine are important metabolic-rate regulators. In fact, thyroxine is one of the most important substances in the body. It influences carbohydrate metabolism as well as protein synthesis and is also involved in cardiovascular, brain, and renal function. Epinephrine and norepinephrine control heart rate, blood flow, and metabolic rate.

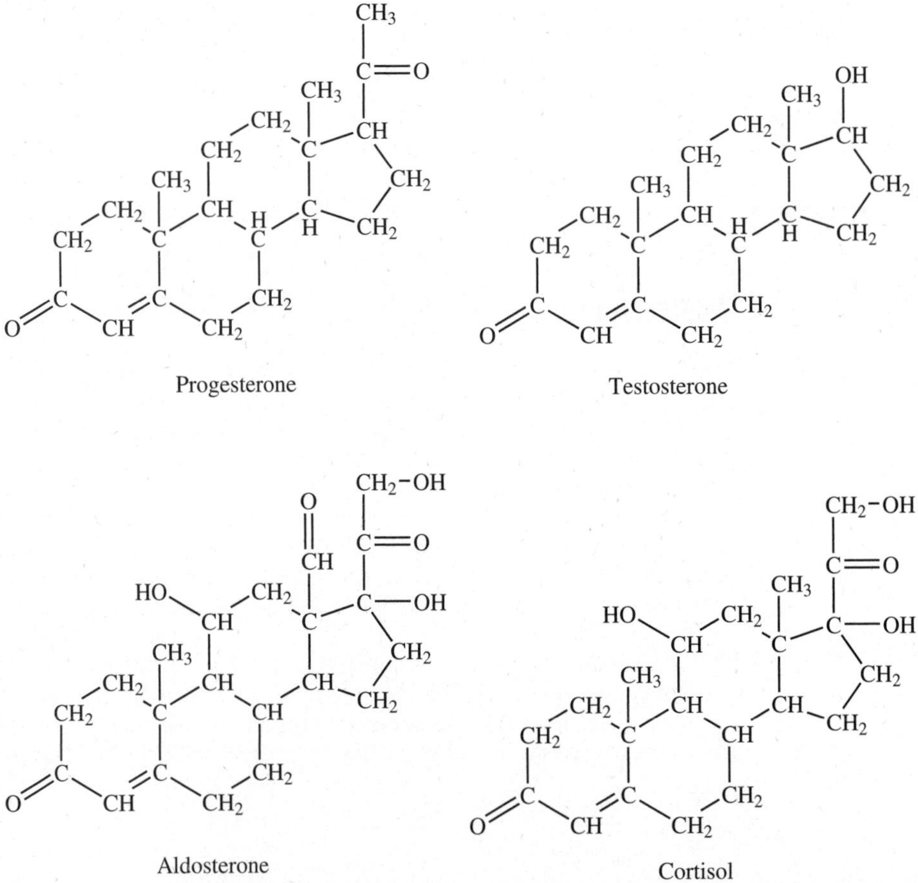

Progesterone

Testosterone

Figure 11-2:
Structures
of proges-
terone (an
estrogen),
testosterone
(an andro-
gen), and
the adrenal
cortical
hormones
aldosterone
and cortisol.

Aldosterone

Cortisol

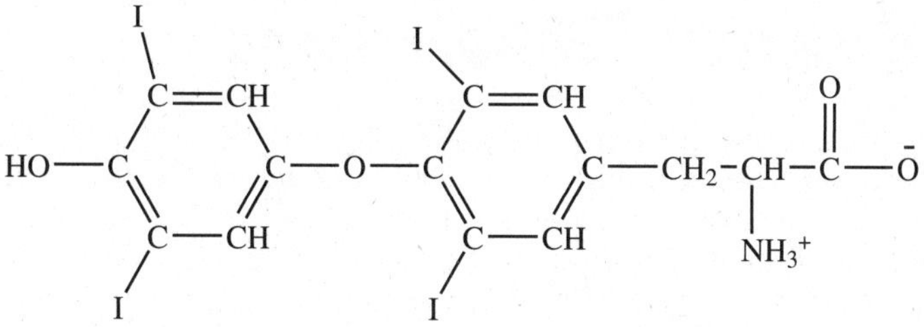

Thyroxine

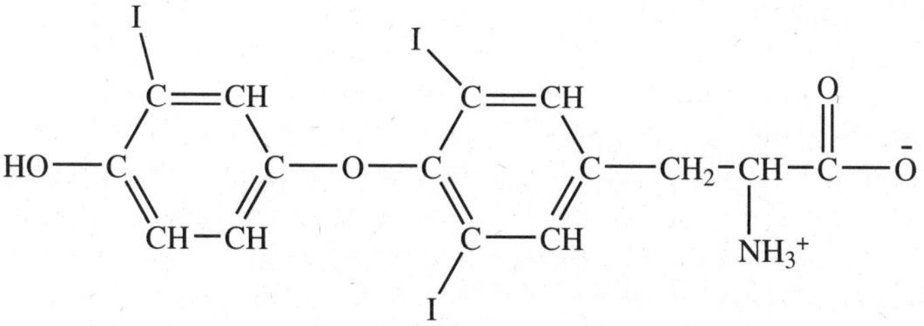

Triiodothyronine

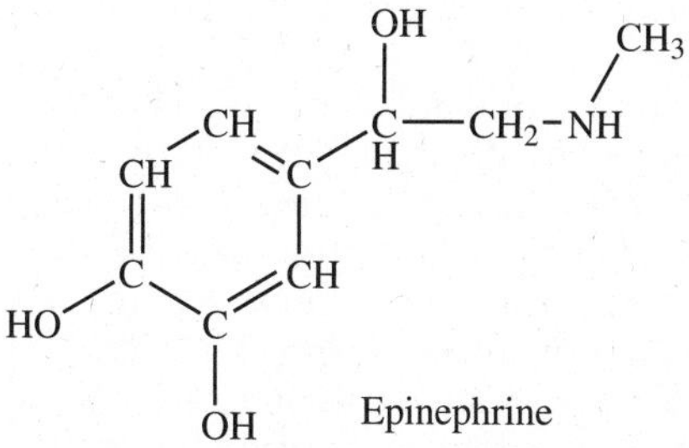

Epinephrine

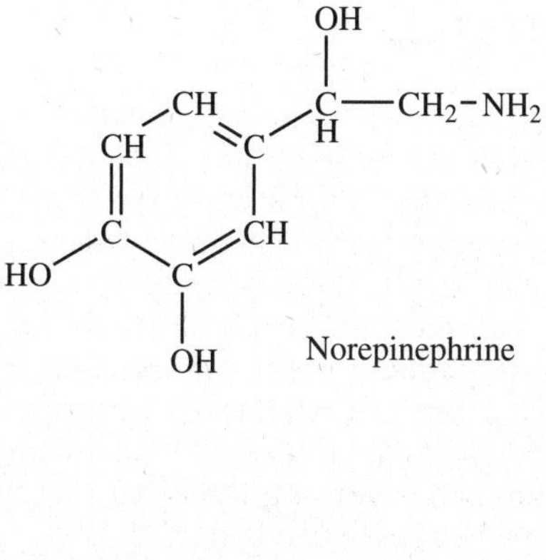

Norepinephrine

Figure 11-3:
Structures
of thyroxine,
triiodothy-
ronine,
epinephrine,
and
norepine-
phrine.

Now and Later: Prohormones

The synthesis of some hormones, like some enzymes, doesn't begin with the production of the molecule in its active form. Instead, a *prohormone* forms, which remains unreactive and dormant until activated — sort of like us in the morning until we get our first cups of coffee. This process allows the body to build a store of a hormone for quick activation. Activating the prohormone requires less time than would the total synthesis of the molecule.

Proinsulin

Proinsulin is an example of a prohormone. *Insulin* is the hormone responsible for controlling blood sugar levels. Too much insulin results in a low blood sugar level *(hypoglycemia)*, whereas too little insulin leads to elevated blood sugar levels *(hyperglycemia)*. Your body needs to have a supply of insulin readily available for when you eat a piece of candy, such as a large chocolate-hazelnut truffle. If all this insulin were already in your bloodstream, upon eating the candy you'd become hypoglycemic. If the insulin weren't present at all, you may become hyperglycemic until your body was able to synthesize sufficient insulin from the individual amino acids. Both hypoglycemia and hyperglycemia can lead to serious medical problems. The presence of a quantity of inactive insulin, ready to jump into action at a moment's notice, is the solution.

Bovine insulin (insulin produced from cows) contains two polypeptide chains, A and B, linked by disulfide linkages, with a total of 51 amino acid residues. Bovine proinsulin has 30 more amino acid residues than insulin does. Proinsulin is a single polypeptide chain with the insulin disulfide linkages. By removing a polypeptide sequence from the central region of this chain (residues 31 to 60), insulin forms. The excised portion originally connected one end of the A chain of insulin to the end of the B chain. The conversion of proinsulin to insulin requires the cleavage of two peptide bonds.

Angiotensinogen

Angiotensinogen is the prohormone of *angiotensin II*, a hormone that signals the adrenal cortex to release aldosterone. (In addition, angiotensin II is the most potent known vasoconstrictor.) The conversion of the prohormone to the hormone requires two steps. The first step uses the enzyme *rennin*. This peptidase, produced in the kidney, specifically cleaves a peptide bond between two leucine residues, the result of which is the *decapeptide angiotensin I.* The second step utilizes the peptidase known as the *angiotensin-converting enzyme.* This enzyme, which occurs primarily in the lungs, cleaves the *C-terminal dipeptide* from angiotensin I to yield angiotensin II. These biochemical reactions can occur very rapidly, ensuring that the hormone can be quickly activated when the body needs it.

Angiotensin and aldosterone's effects on blood pressure

The hormone angiotensin causes blood vessels to constrict, which, in turn, increases the blood pressure. In addition, the protein induces the adrenal cortex to release the hormone aldosterone. Aldosterone increases sodium retention and loss of potassium, which also tends to increase blood pressure.

Together, angiotension and aldosterone are important for the regulation of blood pressure. However, if the levels become too high, they may introduce hypertension. For this reason, many hypertension drugs work specifically with the synthesis or release of one or both of these hormones. The aldosterone level in the blood is expressed as the *plasma aldosterone concentration (PAC)*.

Fight or Flight: Hormone Function

The *endocrine system,* which generates the hormones, consists of a number of apparently unrelated organs: the liver, the ovaries or testes, the thyroid, the pancreas, and a number of other glands — components that are part of a complex, integrated network. A malfunction of one affects others. The hormones epinephrine and norepinephrine are both important to the fight-or-flight response. Both hormones alter many bodily functions.

Opening the letter: Hormonal action

Several mechanisms lead to the regulation of hormones. A *control loop* or *feedback loop* is the simplest. In many cases, one hormone stimulates the production of others so that many actions may occur before some type of control occurs.

Simple control loops

You're familiar with *control loops.* You study for a test but get a so-so grade. So you study harder for the next exam. Your grade provides *feedback,* causing your study habits to (hopefully) change. In the body, a control loop process begins with an external stimulus that signals a gland to generate a hormone. This hormone then influences its target site. Action by the target leads to a change, which signals the gland to stop. The action of the hormone causing the stop signal provides a negative feedback.

An example of this type of loop is the pancreas's production of insulin. The presence of high glucose levels in the bloodstream signals the pancreas to release insulin. The released insulin lowers the glucose level in the bloodstream, and the reduced glucose level signals the pancreas to stop releasing insulin. The low glucose level is the negative feedback. This is a simplification; other factors may trigger the release of insulin. In addition, high glucose levels can trigger other biochemical functions, such as the synthesis of glycogen in the liver.

Hypothalamus-pituitary control

The hypothalamus-pituitary system is a very complex example of the other extreme of hormone control. The hypothalamus and the pituitary glands are in such close proximity that they behave almost as a single unit.

Initially, the central nervous system signals the hypothalamus to release a hormone called a *hormone-releasing factor,* which signals the pituitary. The pituitary, thus signaled, releases another hormone into the bloodstream. This hormone may target a specific organ or signal another part of the endocrine system to secrete yet another hormone. The presence of this final hormone serves as a negative feedback signal to the hypothalamus to stop secreting the hormone-releasing factor to the pituitary. The actions of these hormones influence each other — a series of checks and balances — to make sure you never have too much signal or too little. Again, this is a simplistic view of a complicated system. An analogy might be your parents seeing your so-so exam grade. They freak out and force you to study harder. You're being influenced by an external force — in this case, your parents.

Figure 11-4 gives a more detailed representation of this system. We have introduced some simplification here. For example, the follicle stimulating hormone (FSH) acts in both males and females, but the major effect is in males for testosterone production. Although the pituitary gland is known as the *master gland,* this figure indicates that, in fact, the hypothalamus deserves this honor.

Models of hormonal action

Two models have been proposed to account for the molecular action of hormones. The first is the *second-messenger hypothesis,* which applies primarily to polypeptide and amine hormones. The other, *steroid hormonal action,* applies primarily to steroids. We use a simplistic approach (the KISS rule: Keep It Simple, Silly) for each model to emphasize the basic concepts. The actual processes involve many more changes.

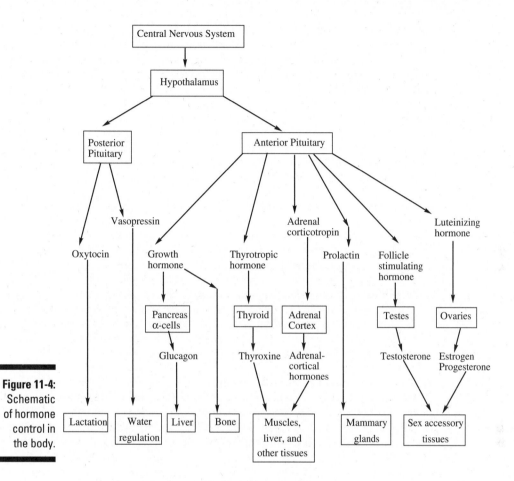

Figure 11-4:
Schematic
of hormone
control in
the body.

The second-messenger model: Like the mail

Studies of the hormonal action of epinephrine (adrenaline) led to the development of the second-messenger model. Later work indicated that the model applies to other hormonal systems as well. In the second-messenger model, a hormone binds to a receptor site on a cell's exterior. This binding induces the release of another agent within the cell. The hormone is the first messenger and the other agent is the second messenger.

For example, the adrenal medulla releases epinephrine, the "fight or flight" hormone, in vertebrate animals. This release initiates a number of responses, including *glycogenolysis* — the breakdown of glycogen. Glycogenolysis releases glucose for use in rapid energy production. As with other hormones, the concentration of hormone that's required is very low. For epinephrine, the concentration is about 10^{-9} M. The released epinephrine acts as the first messenger (the extracellular one). Molecules of epinephrine bind to specific receptor sites on the surface of the target cells — primarily the skeletal muscles and, to a lesser extent, the liver. The binding of epinephrine to the outside of liver cells induces the enzyme *adenylate cyclase,* bound to the cell membrane's interior, to synthesize *cyclic AMP* (see Figure 11-5). Cyclic AMP, or cAMP, is the second messenger (the intracellular one). The second messenger initiates a series of events terminating in the release of glucose (glycogenolysis).

Initially, the cAMP binds to the regulatory subunit of protein kinase, and this activates the membrane-bound enzyme. The released protein kinase then activates phosphorylase kinase. This process requires a calcium ion and ATP. (Muscular action releases a calcium ion, which aids the process.) Phosphorylase kinase, with the aid of ATP and a magnesium ion, converts inactive phosphorylase b to active phosphorylase a. The increased presence of this enzyme accelerates the breakdown of glycogen with the release of D-glucose-1-phosphate. Phosphoglucomutase then converts D-glucose-1-phosphate to D-glucose-6-phosphate. Finally, D-glucose-6-phosphatase catalyzes the loss of the phosphate to release glucose, which may be used in the cell or, more important, may enter the bloodstream. Whew!

The enzyme protein kinase also catalyzes the conversion of glycogen synthase (active) to phospho-glycogen synthase (inactive). Thus, though the level of protein kinase is high, the production of new glycogen ceases. The inhibition of glycogen synthesis also means that more glucose is available for rapid actions, such as running away from an angry pit bull.

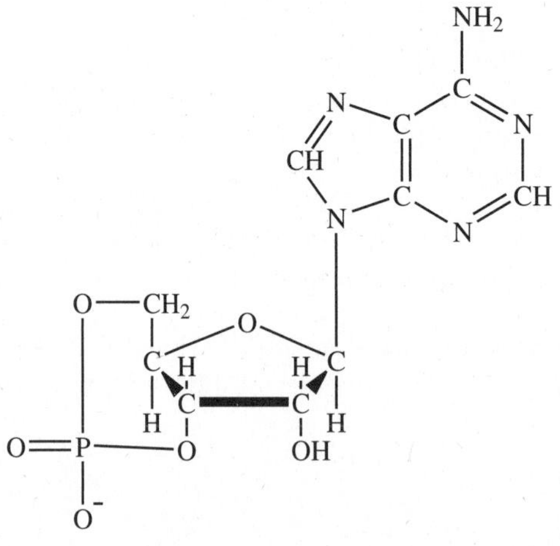

Figure 11-5:
Structure of cyclic AMP.

Amplification

The attachment of a hormone to a receptor site triggers a response. As long as the hormone remains attached, the response will repeat. For example, if the response is the generation of a molecule, then the attached hormone will cause the generation of not just one molecule but of a continual generation of many molecules. This continual generation doesn't stop until the hormone molecule leaves the receptor site. In this way, one hormone molecule may induce the generation of thousands of molecules. This process, known as *amplification*, leads to a significantly greater result than the very low concentration of hormone molecules would indicate.

Steroid hormonal action

Unlike hormones in the second-messenger system, steroid hormones cross the membrane and enter the cell. This mechanism applies to other hormones as well, such as thyroid hormones.

The first system described by this model was the action of *estradiol* on uterine tissue in mammals. Estradiol, an estrogen, passes through the cell's lipid bilayer membrane where it binds to an estrogen-receptor protein. The binding doesn't involve covalent bonding but instead induces a conformational change in the protein. The change in the protein's shape allows it to pass through the "door" into the cell nucleus. The hormone-protein complex then enters the cell nucleus, where it binds to a specific site on a chromosome. This binding to the chromosome stimulates transcription to produce mRNA, and the mRNA exits the nucleus and synthesizes protein molecules through translation.

Three basic factors differentiate the steroid system from the second-messenger system. First, in the steroid system the hormone passes directly through the cell membrane. Second, there's a specific steroid receptor molecule within the *cytosol,* the fluid inside the cell. Finally, the hormone action is at the chromosome level.

Part IV

Bioenergetics and Pathways

The 5th Wave   By Rich Tennant

BEING CHRONICALLY LATE, ANDY ALWAYS
MISSED OUT ON EXAMINING THE REALLY
FUN CELLS

CELL RESEARCH

"Sorry, Andy. All I got left are lung and gut
cells. Take 'em or leave 'em."

In this part . . .

For anyone to do anything requires energy, and in this part we focus on the way life obtains and uses energy. Here we take a gander at energy needs and follow the trail of where that energy goes and why. The main character in this part is your good buddy ATP, and running through this episode is the citric acid (Krebs) cycle. Finally, we tackle nitrogen chemistry.

Chapter 12

Life and Energy

*T*he chapters in Part IV examine *metabolism,* which consists of all the processes involved in maintaining a cell. Metabolism has two components: catabolism and anabolism (neither one of which should be confused with cannibalism). *Catabolism* deals with the breaking down of molecules, whereas *anabolism* deals with the building up of cells. Both processes take place in the cell. All metabolic processes involve energy: They either absorb energy *(endergonic)* or produce it *(exergonic).*

The key energy molecule is *adenosine triphosphate,* abbreviated *ATP,* which forms as a product of the common catabolic pathway.

ATP: The Energy Pony Express

Determining the basic reaction processes involved in the production and use of energy is called *bioenergetics.* This field of study has developed bioenergetic principles that allow scientists to examine energy at the microscopic level.

The typical ATP requirement for an adult is more than 140 pounds per day. However, the amount of ATP present in your body at any one time is only about one-tenth of a pound. Fortunately, ATP is recycled within the body; in fact, each ATP molecule in your body is recycled about 1,400 times each day. Now *that's* effective recycling, and you don't even have to put anything into a blue container!

ATP and free energy

The *free energy content* (G) is the intrinsic energy present in a molecule. In a reaction, the change in this energy is written as $\triangle$G and is equal to the energy of the products minus the energy of the reactants. The value of $\triangle G$ is the key: If a reaction produces energy, $\triangle G$ represents the maximum possible amount of energy that the reaction produces. If a reaction requires energy, $\triangle G$ represents the minimum possible amount of energy that the reaction requires. Reactions that produce energy have a negative value of $\triangle G$ and are *spontaneous.* Reactions that require energy have a positive value of $\triangle G$ and are *nonspontaneous.*

Spontaneity bears no relation to speed. Spontaneous reactions may be very rapid or very slow.

The conditions under which a reaction occurs may alter the value of $\triangle G$. (The "ideal" or standard value of $\triangle G$ is $\triangle G°$.) The formula for modifying the free energy for the equilibrium reaction $A \rightleftarrows B$ is

$$\triangle G = \triangle G° - RT \ln \frac{[B]}{[A]} = \triangle G° - RT \ln K$$

According to this relationship, the free energy change, $\triangle G$, comes from a modification of the standard free energy value. R is the universal gas constant ($8.314 \text{ J} \cdot \text{mol}^{-1}\text{K}^{-1}$ or $1.987 \text{ cal} \cdot \text{mol}^{-1}\text{K}^{-1}$). T is the absolute temperature. K is the equilibrium constant found by dividing the concentration of the product, [B], by the concentration of the reactant, [A].

In many bioenergetic studies, *calories* are the unit instead of joules (J). The relationship is 1 calorie = 4.184 J (exactly) or 1 kilocalorie = 4.184 kJ.

In research, it's often better to use $\triangle G°'$. This modification of $\triangle G°$ stems from the use of the biologically more realistic value of pH = 7 ([H$^+$] = 10^{-7} M) instead of the standard pH = 0 ([H$^+$] = 1 M). Table 12-1 shows some relationships between K and $\triangle G°'$.

Table 12-1	Relationships between $\triangle G°'$ and K
$\triangle G°'$ *kJ • mol^{-1}*	*K*
−17.1	1,000
−11.4	100
−5.7	10
0	1
5.7	0.1
11.4	0.01
17.1	0.001

Table 12-1 shows that the larger K is, the more *exergonic* (spontaneous) the reaction. For example, if K = 1,000, the concentration of the product, [B], is 1,000 times that of the reactant, [A], and 17 kJ per mole is released. Remember that in biological systems, you must take variations in [A] and [B] into account in addition to $\triangle G^{\circ\prime}$. For example, increasing the reactant concentration promotes the reaction, whereas increasing the product concentration inhibits the reaction.

ATP as an energy transporter

Cells utilize exergonic processes to provide the energy necessary for life processes, and the key supplier of this energy is ATP (see Figure 12-1). ATP supplies the energy required to force endergonic (nonspontaneous) reactions to take place to provide mechanical energy (muscle movement), light energy (in fireflies), and heat energy (to maintain body temperature).

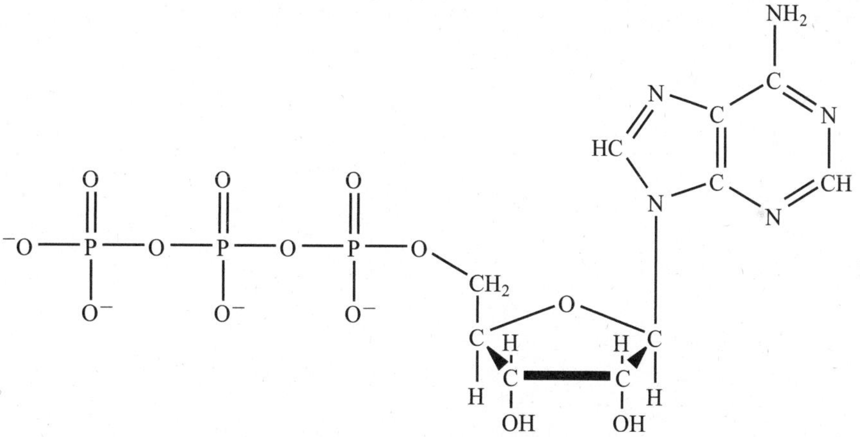

Figure 12-1:
Structure
of ATP.

Adenosine triphosphate (ATP)

Hydrolysis of the terminal phosphate of ATP yields *ADP* and *inorganic phosphate,* indicated as P_i. Figure 12-2 shows the structure of ADP. This hydrolysis releases 30.5 kJ • mol^{-1}.

Concentration variations lead to changes, usually minor, in energy.

Hydrolysis of the terminal phosphate of ADP yields *AMP* and inorganic phosphate, indicated as P_i. Figure 12-3 shows the structure of AMP. This hydrolysis also releases 30.5 kJ • mol^{-1}. (This reaction is of less biological importance than the ATP to ADP hydrolysis.)

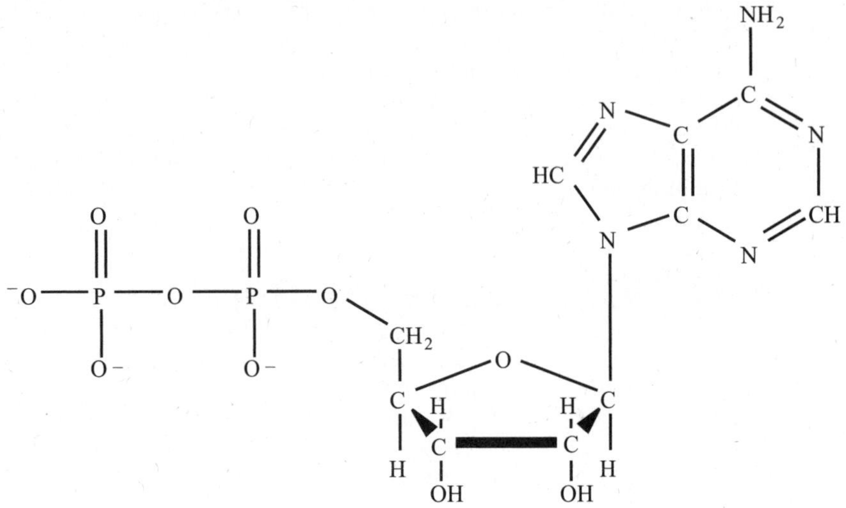

Figure 12-2:
Structure
of ADP.

Adenosine diphosphate (ADP)

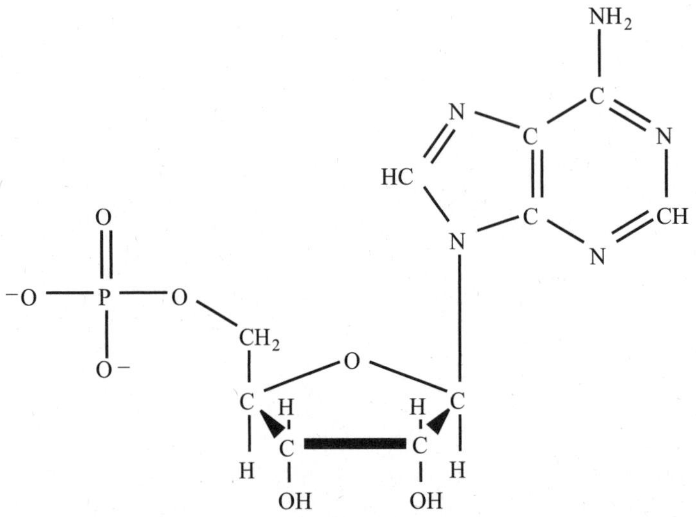

Figure 12-3:
Structure
of AMP.

Adenosine monophosphate (AMP)

Going directly from ATP to AMP is also possible, cleaving a *pyrophosphate,* $P_2O_7^{4-}$, from the phosphate chain. Biochemists use PP_i to indicate a pyrophosphate. This reaction furnishes slightly more energy than a simple hydrolysis to release P_i (about 33.5 kJ • mol^{-1}). Under physiological conditions, the phosphate portions of ATP and ADP form a complex with magnesium ions. In certain circumstances, manganese (II) ions, Mn^{2+}, may take the place of Mg^{2+} ions. Figure 12-4 depicts the magnesium complexes with ATP and ADP.

The removal of the last phosphate involves the loss of the least amount of energy (14.2 kJ • mol^{-1}). This hydrolysis involves the cleavage of an ester bond instead of an anhydride bond. In general, the hydrolysis of an ester bond involves less than half the energy of the hydrolysis of an anhydride bond.

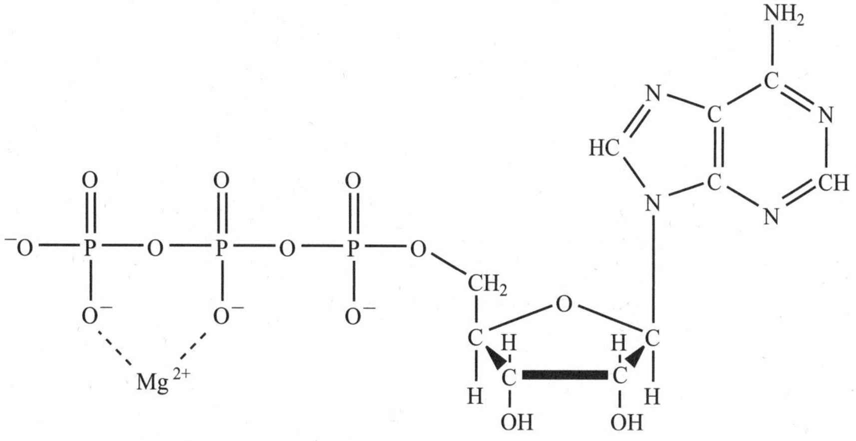

Adenosine triphosphate (ATP)–Mg^{2+}

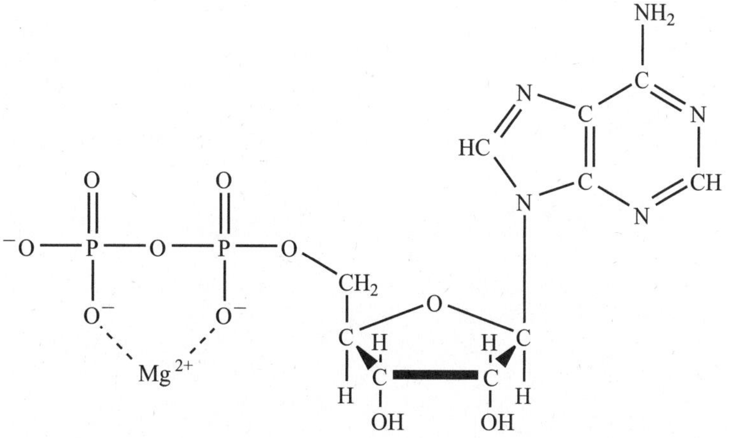

Figure 12-4:
Magnesium complexes with ATP and ADP.

Adenosine diphosphate (ADP)–Mg^{2+}

It's Relative: Molecules Related to ATP

A few other biomolecules can provide energy equivalent to the energy that comes from the hydrolysis of ATP. *GTP* is an example of such a molecule. A few molecules also supply *more* energy. Table 12-2 compares some of the high-energy molecules to ATP, and Figure 12-5 shows their structures.

Table 12-2	Energy Released ($\triangle G°'$) by Some High-Energy Molecules Related to ATP
Biomolecule	*Energy Released (kJ • mol^{-1})*
ATP	30.5
Phosphoarginine	32.2
Acetyl phosphate	43.3
Phosphocreatine	43.3
1,3-Bisphosphoglycerate	49.6
Phosphoenolpyruvate	62.2

Phosphoenolpyruvate, 1,3-bisphosphoglycerate, and acetyl phosphate are important for the transfer and conservation of chemical energy. Phosphoarginine and phosphocreatine are important molecules for storing metabolic energy. Phosphocreatine is stored in muscles and can be quickly converted to ATP to give energy for muscle contraction. Production of phosphocreatine occurs when ATP concentration is high; high ATP concentration is needed to overcome the energy deficit of 12.8 kJ • mol^{-1}. The reverse, phosphate transfer to form ATP from ADP, occurs at low ATP concentrations. Phosphoarginine behaves similarly in certain invertebrates (talk about strong mussels!).

The nucleoside triphosphate family

The predominant energy transfer molecule, as we have been saying, is ATP. But other nucleoside triphosphates, such as CTP, GTP, TTP, and UTP, may also serve this energy transfer function. These five molecules also supply part of the energy necessary for DNA and RNA synthesis. All the nucleoside triphosphates have about the same energy yield. (Note that ATP is necessary for the synthesis of the remaining nucleoside triphosphates.)

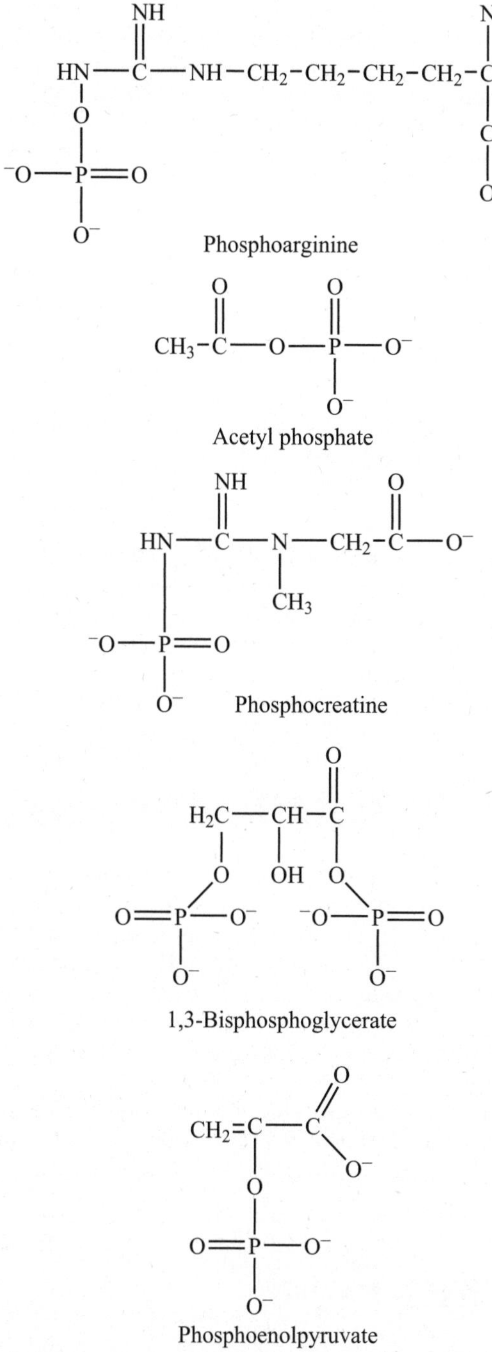

Figure 12-5:
Structures
of some
high-energy
molecules.

The biosynthesis of the ribonucleoside triphosphates, in general NTP, begins with the production of the appropriate monophosphate, NMP. The stepwise addition of the next two phosphate groups requires two enzymes of low specificity. These enzymes are *nucleoside monophosphate kinase* and *nucleoside diphosphate kinase.* (The term *kinase* refers to a transferase enzyme that transfers a phosphate group of a nucleoside triphosphate.) Figure 12-6 shows the general reactions.

Figure 12-6: Two of the reactions catalyzed by the kinase enzymes.

Nucleoside monophosphate kinase

$$NMP \quad + \quad ATP \quad \rightleftharpoons \quad NDP \quad + \quad ADP$$

Nucleoside diphosphate kinase

$$NDP \quad + \quad ATP \quad \rightleftharpoons \quad NTP \quad + \quad ADP$$

The formation of the deoxyribonucleoside triphosphates, dNTP, follows two different paths. In one path, a multi-enzyme system converts the appropriate nucleoside diphosphate to the corresponding deoxyribonucleoside diphosphate. Then nucleoside diphosphate kinase catalyzes the formation of the deoxyribonucleoside triphosphate. The other path occurs in certain microorganisms in which there's a direct conversion of NTP to dNTP.

As easy as 1, 2, 3: AMP, ADP, and ATP

It's possible to hydrolyze ATP either to ADP plus phosphate, P_i, or to AMP plus pyrophosphate, PP_i. (The pyrophosphate undergoes further hydrolysis to two phosphates, $2\,P_i$.) ADP and P_i are the immediate precursors for the reformation of ATP. To produce ATP starting with AMP, the enzyme *adenylate kinase* is utilized. This enzyme catalyzes the transfer of a phosphate group from an ATP to an ADP. This reaction results in the formation of two ADP molecules. (Adenylate kinase also catalyzes the reverse reaction.)

The easy transfer of phosphate groups among nucleotides creates a metabolic network for the transfer of energy. The key to this network is the intercellular production of ATP.

Where It All Comes From

One of the purposes of the food you eat, of course, is to supply energy, with carbohydrates and fats being the major energy sources. Digestion breaks polysaccharides into glucose and other monosaccharides, whereas fats are broken into glycerol and fatty acids. Catabolism converts these energy sources primarily to ATP. Proteins are broken into amino acids, which don't usually serve as energy sources. (We explain the details of these reactions later in this book.) Glucose produces 36 ATP molecules, an average of 6 ATPs per carbon. Table 12-3 shows the step-by-step energy change for glucose. Other carbohydrates give a similar yield.

Table 12-3	ATP Yield for Each Step in the Metabolism of Glucose
Chemical Steps	**Number of ATP Molecules Produced**
Activation (conversion of glucose to 1,6-fructose diphosphate)	−2
Oxidative phosphorylation 2(glyceraldehyde 3-phosphate → 1,3-diphosphoglycerate), producing 2 NADH + H$^+$ in cytosol	4
Dephosphorylation 2(1,3-diphosphoglycerate → pyruvate)	4
Oxidative decarboxylation 2(pyruvate → acetyl CoA), producing 2 NADH + H$^+$ in mitochondria	6
Oxidation of two C$_2$ fragments in citric acid and oxidative phosphorylation common pathway, producing 12 ATP for each C$_2$ fragment	24
Total	36

What happens if you stop eating?

Starvation is the total deprivation of food. Here's what happens during starvation: Initially, the body utilizes its glycogen reserves. Then it moves on to its fat reserves; the first ones are those around the heart and kidneys. Finally, the body relies on the reserves found in muscles and the bone marrow.

Early in a total fast, the body metabolizes protein at a rapid rate. The amino acids are converted to glucose because the brain prefers glucose. These proteins come from the skeletal muscles,

blood plasma, and other sources in a process that produces a quantity of nitrogen-containing products, which need to be excreted. Excretion requires large quantities of water, and the resulting loss of water may lead to death by dehydration.

If the starvation continues, the brain chemistry adjusts to accept fatty acid metabolites, which uses the last of the fat reserves. Finally, the body resorts to structural proteins, systems begin to fail rapidly, and death follows quickly.

Each fat molecule hydrolyzes to a glycerol and three fatty acid molecules. Glycerol produces 20 ATPs per molecule. The energy production from a fatty acid varies with the identity of the particular acid. Stearic acid, $C_{18}H_{36}O_2$, produces a total of 146 ATPs per molecule — an average of 8.1 ATPs per carbon. Table 12-4 shows the step-by-step energy change for stearic acid. Other fatty acids give a similar yield.

Table 12-4	ATP Yield for Each Step in the Metabolism of Stearic Acid	
Chemical Steps	*Number of Times the Step Happens*	*Number of ATP Molecules Produced*
Activation (stearic acid → stearyl CoA)	Once	−2
Dehydrogenation (acyl CoA → transenoyl CoA), producing $FADH_2$	8 times	16
Dehydrogenation (hydroxyacyl CoA → keto acyl CoA), producing NADH + H⁺	8 times	24
C_2 fragment (acetyl CoA → common catabolic pathway), producing 12 ATP per C_2 fragment	9 times	108
Total		146

Chapter 13

ATP: The Body's Monetary System

*1*n this chapter we examine a number of general processes that either produce or consume energy. Breaking down molecules often produces energy. The breakdown of one molecule is often coupled with the synthesis of another, and this other synthesized molecule is often adenosine triphosphate, or ATP. *Catabolism* is the breaking down of molecules to provide energy. *Anabolism* is the building of molecules. These two processes combine to give *metabolism,* which comprises the reactions in biological systems.

As you can see in Chapter 12, the "currency" in biological systems is ATP. There are other energy-containing molecules, but the rate of exchange to ATP is the reference. The breakdown of certain molecules produces the currency of ATP, and a cost is involved in the synthesis of other molecules. Polysaccharides and fats are like "banks" that store energy for later use.

Metabolism 1: Glycolysis

The *Embden-Meyerhof pathway* (sounds like a German highway), or *glycolysis,* is a primitive means of extracting energy from organic molecules. The process is *anaerobic* (without oxygen) and converts glucose to two lactic acid molecules. Nearly all forms of life, whether a human or a jellyfish, utilize glycolysis. All carbohydrates follow this pathway. *Aerobic* (utilizing oxygen) processing of carbohydrates uses pyruvate derived from glycolysis. (Alcoholic fermentation also produces pyruvate from glucose. The glucose is converted to two ethanol molecules and two CO_2 molecules.) Glycolysis is a two-part process, which we label Phase I and Phase II. Figures 13-1 and 13-2 help illustrate the upcoming, ahem, rather *involved* discussion. You may want to refer back to these figures as you read.

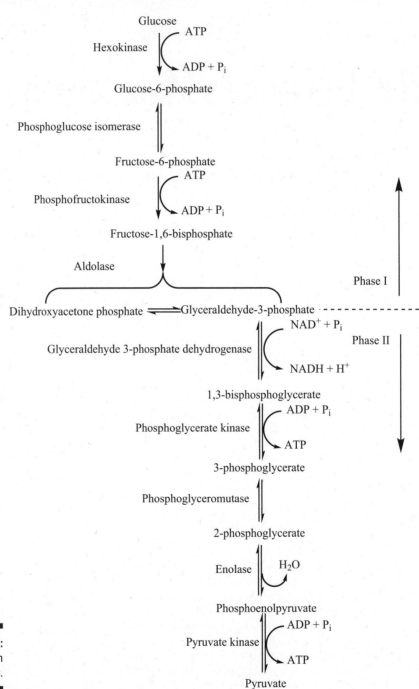

Figure 13-1:
Steps in
glycolysis.

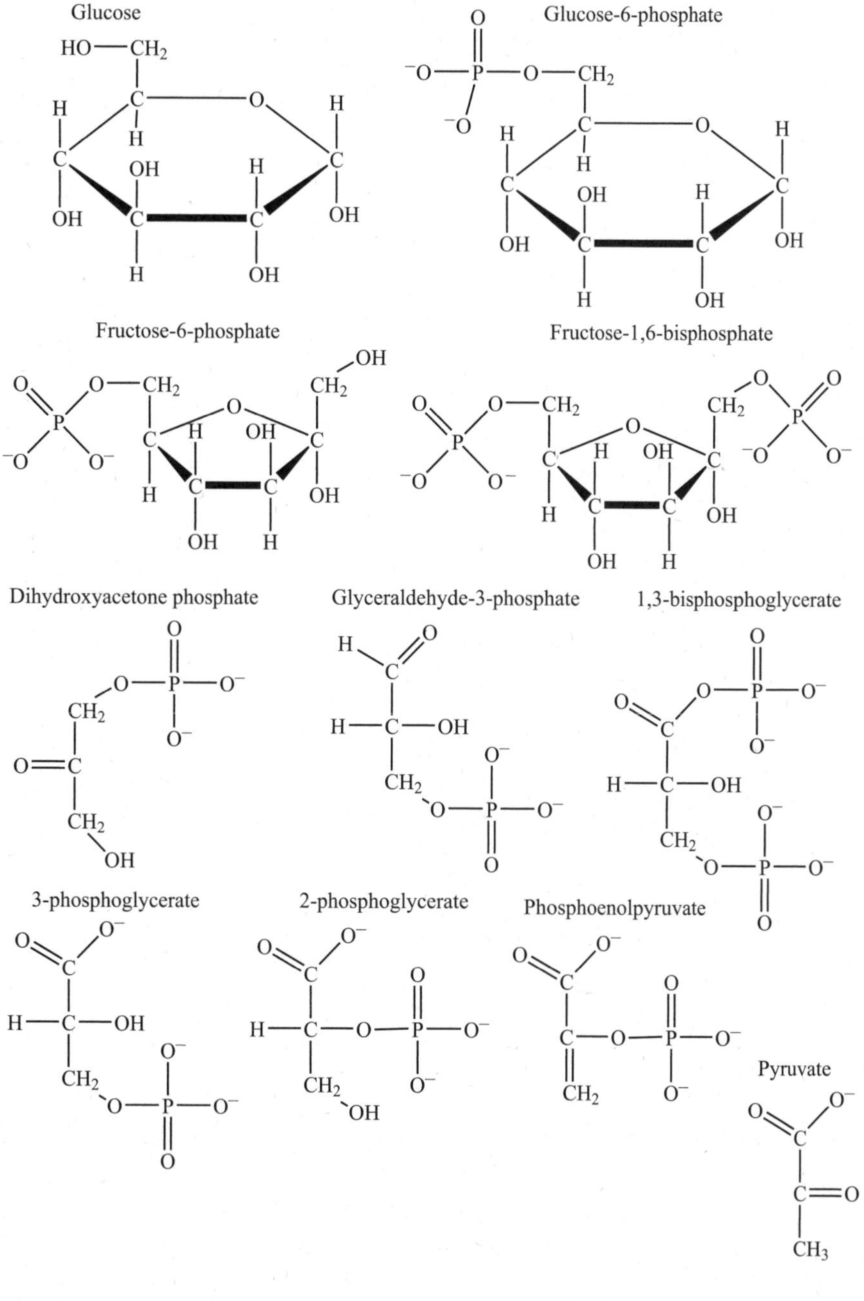

Figure 13-2:
Molecules
involved in
glycolysis.

Glucose: Where it all starts

Glycolysis occurs in two phases: Phase I and Phase II.

Phase 1

As glucose enters the cell, it undergoes immediate phosphorylation to glucose-6-phosphate — the first step in Phase I. The phosphate comes from ATP, and the enzyme hexokinase, with the aid of Mg^{2+}, catalyzes the transfer. Thus, the first step in the production of energy requires an investment of energy, which is necessary to activate the glucose in a reaction that isn't easy to reverse. In addition, the presence of the charged phosphate group makes it difficult for this and other intermediates to diffuse out of the cell.

The enzyme phosphoglucose isomerase then catalyzes the isomerization of glucose-6-phosphate to fructose-6-phosphate. This results in a compound with a primary alcohol group, which is easier to phosphorylate than the hemiacetal originally present. Fructose-6-phosphate then reacts with another molecule of ATP to form fructose-1,6-bisphosphate. The enzyme for this step is phosphofructokinase — try saying that ten times fast! — and this enzyme requires Mg^{2+} to be active. ATP inhibits this enzyme, whereas AMP activates it. This is the major regulatory step in glycolysis.

Aldolase enzymatically cleaves the fructose-1,6-bisphosphate into two triose phosphates. These triose phosphates are dihydroxyacetone phosphate and glyceraldehyde-3-phosphate. The dihydroxyacetone phosphate isomerizes to glyceraldehyde-3-phosphate to complete Phase I. Triose phosphate isomerase catalyzes this isomerization. (You see why we suggested following along with Figures 13-1 and 13-2?)

The net result of Phase I is the formation of two molecules of glyceraldehyde-3-phosphate, which costs two ATP molecules and produces no energy.

Phase 11

Phase II begins with the simultaneous phosphorylation and oxidation of glyceraldehyde-3-phosphate to form 1,3-bisphosphoglycerate. Glyceraldehyde-3-phosphate dehydrogenase catalyzes this conversion. Inorganic phosphate is the source of the phosphate. NAD^+ is the coenzyme and oxidizing agent. NAD^+ reduces to NADH.

There's a high-energy acyl phosphate bond present in 1,3-bisphosphoglycerate. Phosphoglycerate kinase, in the presence of Mg^{2+}, catalyzes the direct transfer of phosphate from 1,3-bisphosphoglycerate to ADP. This results in the formation of ATP and 3-phosphoglycerate. Because the formation of ATP involves direct phosphate transfer, this process is called *substrate-level phosphorylation* to avoid confusion with *oxidative phosphorylation* (discussed later). Phosphoglyceromutase then catalyzes the transfer of a phosphate group from C-3 to C-2, thus converting 3-phosphoglycerate to 2-phosphoglycerate.

After that, dehydration occurs to form phosphoenolpyruvate (PEP), which contains a high-energy phosphate bond. The enzyme that catalyzes the reaction is enolase.

The final, irreversible step is a second substrate-level phosphorylation. Here, an ADP molecule receives a phosphate group from the PEP. The enzyme pyruvate kinase is necessary for this step. This enzyme requires not only Mg^{2+}, but also K^+. Pyruvate is the other product. Whew! Take a deep breath before going on.

During Phase II, two molecules of glyceraldehyde-3-phosphate (from Phase I) form two molecules of pyruvate with the formation of four molecules of ATP and two molecules of NADH.

The pyruvate produced by glycolysis has several fates. When plenty of oxygen is available, the pyruvate enters the Krebs cycle, the electron transport chain, and oxidative phosphorylation pathways as acetyl-CoA. This results in the production of more ATP and the total conversion to CO_2. If oxygen is lacking, vertebrates (you included) convert pyruvate to a related substance, lactate. Other organisms, such as yeast, convert pyruvate to ethanol and CO_2, and that's how beer is made. (Blessed yeasts!) These latter two possible fates yield less energy than the oxygen-rich fate.

Releasing the power: Energy efficiency

Glycolysis is the initial conversion of carbohydrate to energy. After that is the production of two ATP molecules, two NADH molecules, and two pyruvate molecules. The energy content of the ATP molecules is only 2 percent of the total energy present in each glucose molecule, a number that shows the relative inefficiency of anaerobic energy production. Fortunately, the pyruvate molecules undergo further aerobic oxidation to increase this energy yield. The total energy output of anaerobic and aerobic oxidation of glucose is 36 ATP molecules, which accounts for about 30 percent of the total energy present in glucose. Much of the remaining energy is available as heat for warmblooded animals.

Going in reverse: Gluconeogenesis

Gluconeogenesis is a series of reactions that generate glucose from non-carbohydrate sources. This pathway is necessary when the supply of carbohydrates is inadequate (something that's rare in our lives). The noncarbohydrate sources include lactate, pyruvate, some amino acids, and glycerol. In many ways, gluconeogenesis is the reverse of glycolysis. Figure 13-3 summarizes the steps of gluconeogenesis. (The formation of glucose in plants utilizes the process of photosynthesis.)

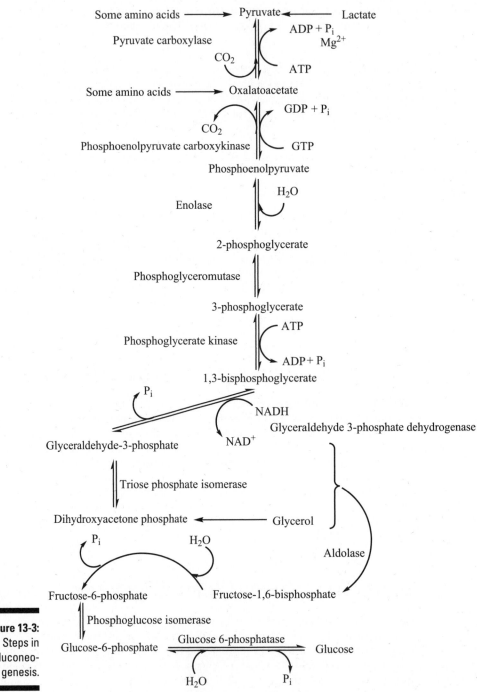

Figure 13-3:
Steps in
gluconeo-
genesis.

The presence of many of the same intermediates enables the use of many of the same enzymes in both glycolysis and gluconeogenesis. The differences (four enzymes) between the two systems allow regulation, so that the processes don't cancel each other. Regulation is also possible by isolating the two pathways in different organs. Other carbohydrates may also form.

Alcoholic fermentation: We'll drink to that

Under anaerobic conditions, yeast and other organisms convert pyruvate to ethanol and carbon dioxide. (The carbon dioxide has a very useful purpose — the bubbles in beer and champagne come from this fermentation reaction. Hence the term "bubbly" for champagne.) This process is accompanied by the oxidation of NADH to NAD^+. The NAD^+ is used in glycolysis. This process yields a net generation of two ATP molecules.

The first step in alcoholic fermentation is the decarboxylation of pyruvate to carbon dioxide and acetaldehyde. The enzyme pyruvate decarboxylase, along with the cofactors Mg^{2+} and TPP (thiamine pyrophosphate), catalyzes this step. The enzyme alcohol dehydrogenase, along with the coenzyme NADH, catalyzes the conversion of acetaldehyde to ethanol. Makes you really appreciate that shot of tequila, doesn't it? Figure 13-4 summarizes these steps.

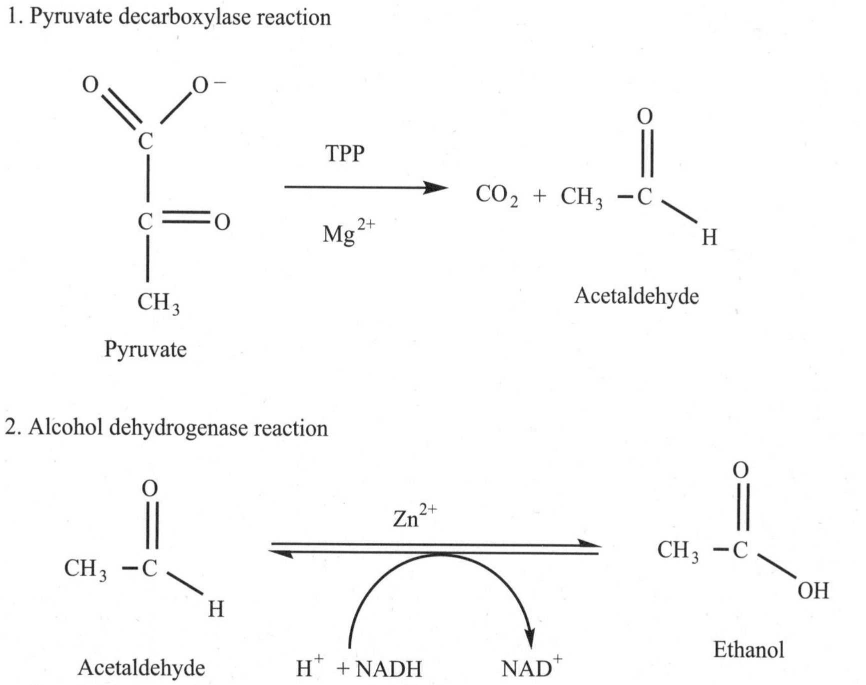

1. Pyruvate decarboxylase reaction

2. Alcohol dehydrogenase reaction

Figure 13-4: Steps in alcoholic fermentation.

Metabolism II: Citric Acid (Krebs) Cycle

The *citric acid cycle* and *oxidative phosphorylation* are the aerobic processes of catabolism that produce energy (ATP). The citric acid cycle is also known as the *Krebs cycle* or the *tricarboxylic acid cycle* (TCA). The primary entry molecule for this series of reactions is acetyl-CoA (short for acetyl-coenzyme A). The sources of acetyl-CoA are pyruvate from glycolysis, certain amino acids, or the fatty acids present in fats. Figure 13-5 shows the structure of acetyl-CoA. *Note:* These processes take place in the *mitochondria,* the energy factories of the cell.

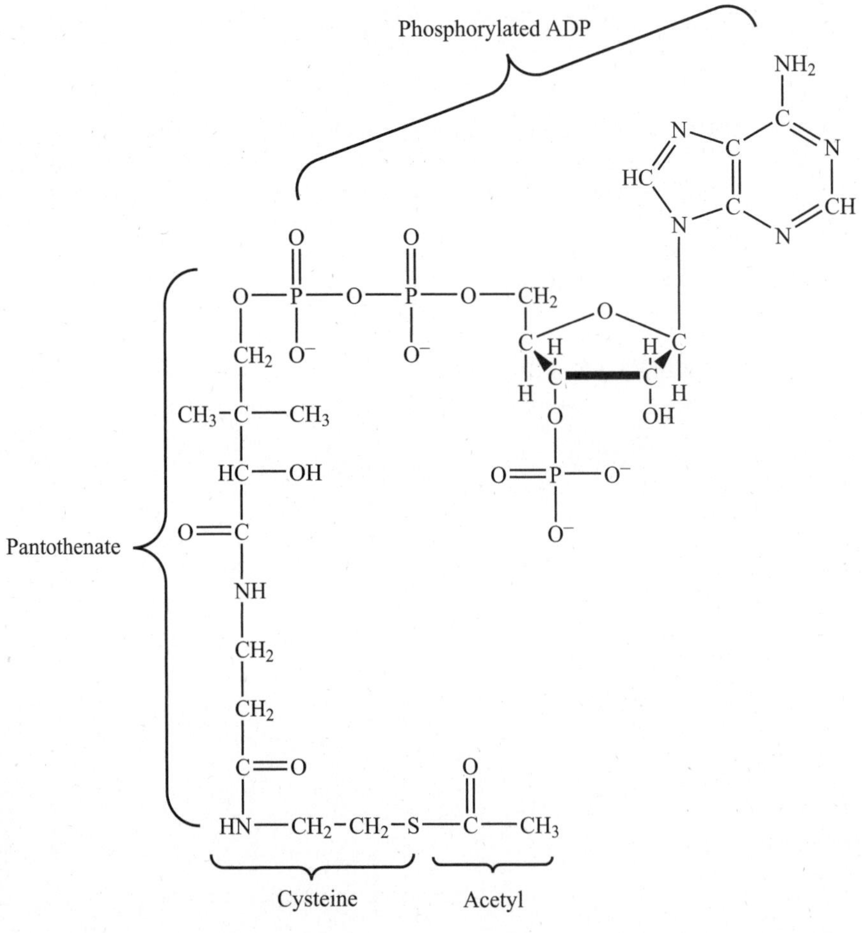

Figure 13-5:
Structure of
acetyl-CoA.

In addition to being an energy source, acetyl-CoA is the starting material for the synthesis of a number of biomolecules. In the next few sections, we discuss the citric acid cycle. The general cycle is shown in Figure 13-6, and the structures are shown in Figure 13-7.

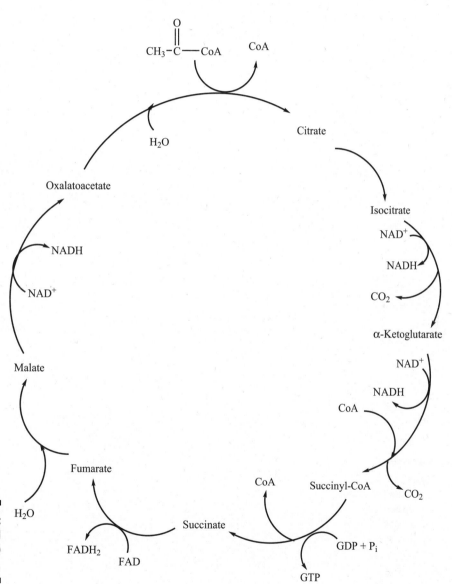

Figure 13-6: Citric acid (Krebs) cycle.

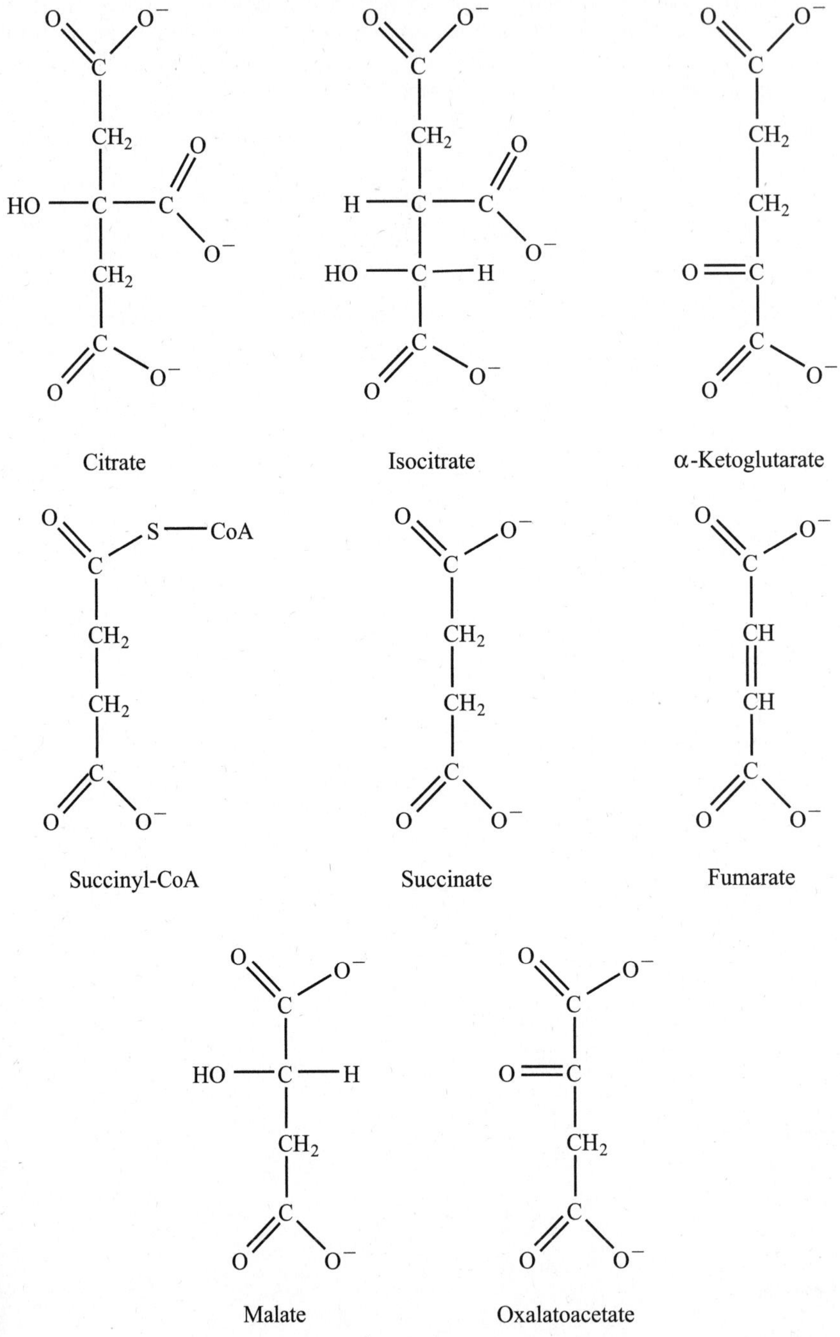

Figure 13-7:
Structures
of mol-
ecules
involved in
the citric
acid (Krebs)
cycle.

Let's get started: Synthesis of acetyl-CoA

The synthesis of acetyl-CoA is a multistep process. These steps are coupled to preserve the free energy produced by the decarboxylation. In the first step, pyruvate combines with TPP (thiamine pyrophosphate) and undergoes decarboxylation. (Figure 13-8 shows a simplified version of these steps.) The pyruvate dehydrogenase component of the multi-enzyme complex catalyzes this step. During the second step, the TPP derivative undergoes oxidation, which yields an acetyl group. This acetyl group transfers to lipoamide. In this reaction, the oxidant is the disulfide group of lipoamide, and acetyllipoamide forms. The pyruvate dehydrogenase component also catalyzes this reaction. In the final step, the acetyl group of acetyllipoamide transfers to CoA to form acetyl-CoA. The catalyst for this reaction is dihydrolipoyl transacetylase (another component of the pyruvate dehydrogenase complex).

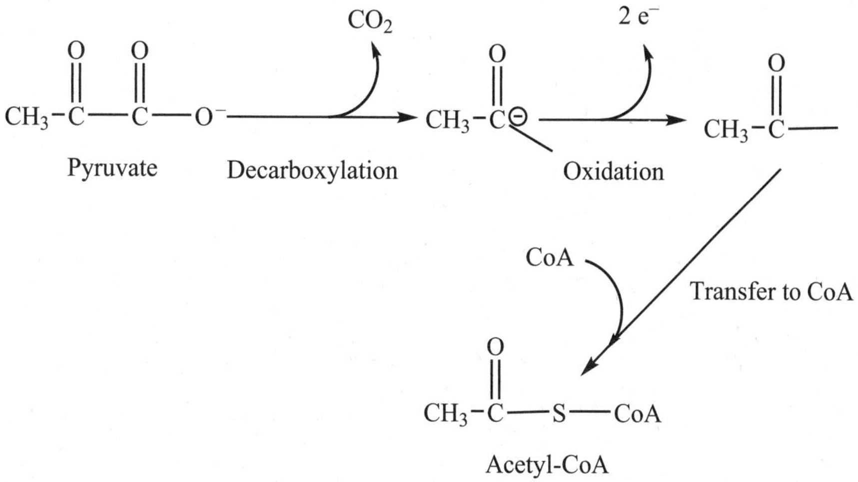

Figure 13-8: Simplified scheme for the formation of acetyl-CoA.

However, the process doesn't end with the formation of acetyl-CoA. The oxidized form of lipoamide must be regenerated. The enzyme dihydrolipoyl dehydrogenase catalyzes this step. The two electrons from the oxidation transfer to FAD and then to NAD$^+$. Figure 13-9 shows some of the important intermediates generated in these steps.

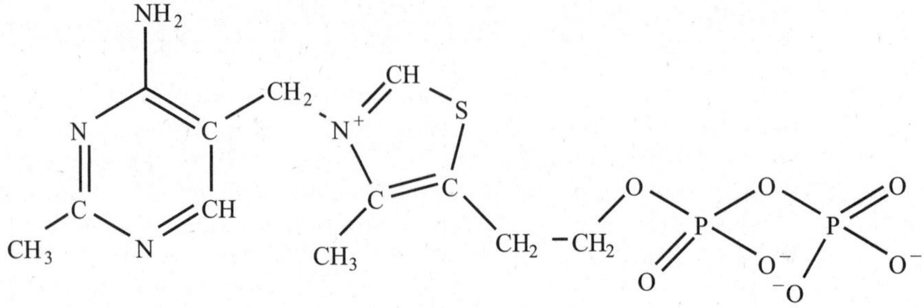

Thiamine pyrophosphate (TPP)

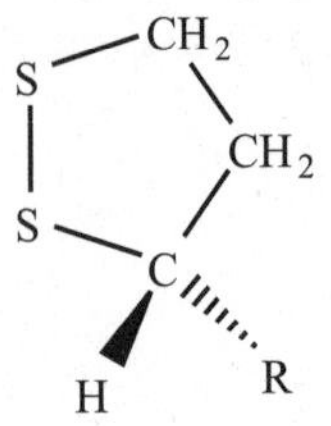

Figure 13-9:
Structures
of TPP,
lipoamide,
and
acetyllipo-
amide.

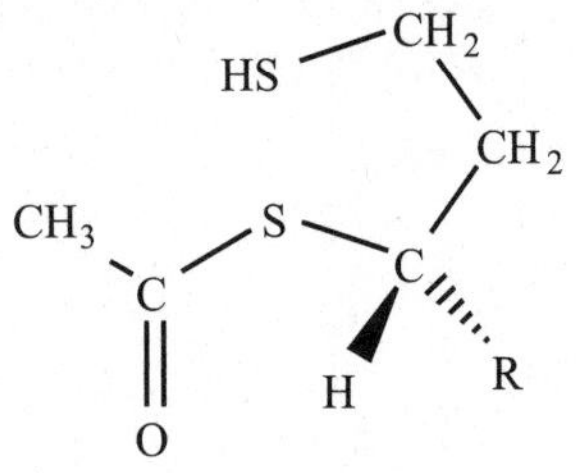

Lipoamide

Acetyllipoamide

Three's a crowd: Tricarboxylic acids

When acetyl-CoA enters the citric acid cycle, it interacts, in the presence of citrate synthase, with oxaloacetate. This interaction results in the transfer of the acetyl group to the oxaloacetate to form citrate. The hydrolysis of the thioester linkage of the acetyl-CoA releases a large amount of energy.

The enzyme aconitase, with Fe^{2+} as a cofactor, catalyzes the isomerization of citrate to isocitrate. For a time, cis-aconitase, derived from aconitase, was thought to be a part of the citric acid cycle. However, the structure of cis-aconitate is related to an intermediate in the formation of isocitrate and is part of the citric acid cycle. Figure 13-10 shows the structure of cis-aconitate.

Oxidative decarboxylation

The next step is the conversion of isocitrate to α-ketoglutarate. The molecule passes through the intermediate oxalosuccinate. The isocitrate binds to the enzyme isocitrate dehydrogenase. During this process, the coenzyme NAD^+ undergoes reduction. Both ATP and NADH are negative factors in the allosteric regulation of isocitrate dehydrogenase, whereas ADP is a positive factor. This is an important mechanism to control the production of ATP.

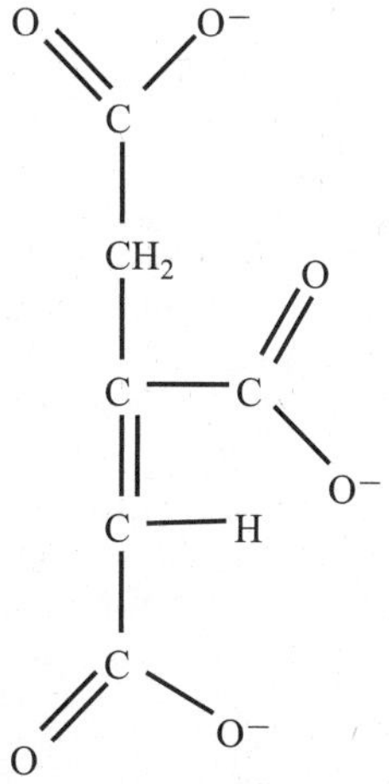

Figure 13-10: Structure of cis-aconitate.

Production of succinate and GTP

The conversion of α-ketoglutarate to succinate requires two steps. The α-ketoglutarate must bind α-ketoglutarate dehydrogenase to form a complex. This reaction requires the same cofactors as needed for the formation of acetyl-CoA. The result of this reaction is the elimination of carbon dioxide and the formation of succinyl-CoA. This process is irreversible under physiological conditions.

In the second step, succinyl-CoA separates to form succinate and release energy, which is harnessed by the conversion of GDP to GTP. This substrate-level phosphorylation is catalyzed by succinyl-CoA synthetase. (GTP contains about the same energy as ATP and can substitute for ATP.)

Oxaloacetate regeneration

The regeneration of oxaloacetate completes the cycle, requiring three reactions which, together, convert a methylene to a carbonyl group. First, a hydrogen atom is removed from each of two adjacent carbon atoms, resulting in the formation of a double bond. Next, a water molecule adds to the double bond. Finally, the removal of two hydrogen atoms yields the appropriate α-keto group. Succinate dehydrogenase catalyzes the first of these reactions. The prosthetic group, FAD, accepts the two hydrogen atoms by covalently binding to the enzyme. Fumarase catalyzes the next step. The final oxidation utilizes the enzyme malate dehydrogenase with the coenzyme NAD^+. The oxaloacetate is now ready to begin the cycle again.

Amino acids as energy sources

The process of using amino acids as energy sources begins with the removal of the amino group. This usually occurs through *transamination,* which is the transfer of an amino group from one molecule to another. All amino acids other than threonine, proline, and lysine undergo this process. Usually, the amino group transfers to the keto carbon of α-ketoglutarate, oxalatoacetate, or pyruvate to form glutamate, aspartate, or alanine, respectively. Specific transaminases are necessary, and the coenzyme pyridoxal phosphate catalyzes this process. A second transamination is involved in the process of transforming aspartate and alanine to glutamate.

Oxidative deamination converts glutamate to α-ketoglutarate. This process, which occurs primarily in the liver, releases an ammonium ion. The reverse reaction, glutamate synthesis, is one of the few reactions that occur in animals in which inorganic nitrogen is converted into organic nitrogen. The ammonium ion resulting from oxidative deamination may enter one or more biosynthetic pathways or the urea cycle. Most vertebrates convert the ammonium ion to urea, which is excreted in the urine. Most marine organisms, including fish, eliminate ammonia directly, whereas birds, insects, and reptiles convert the ammonium ion to uric acid.

The products of transamination, oxidative deamination, and further modification of the remaining portion of the amino acid produce one of the intermediates in glycolysis or the citric acid cycle. This is the fate of all the amino acids — some of the amino acids go through one intermediate, whereas others require more intermediates. Figure 13-11 shows where each of the amino acids enters glycolysis; some of the amino acids have more than one entry point.

Although carbohydrates are the most readily available energy source, amino acids can serve as energy sources in some situations. This is important for carnivores (like humans), who live on a high-protein diet. The utilization of amino acids as energy sources is also important during hypoglycemia, fasting, and starvation.

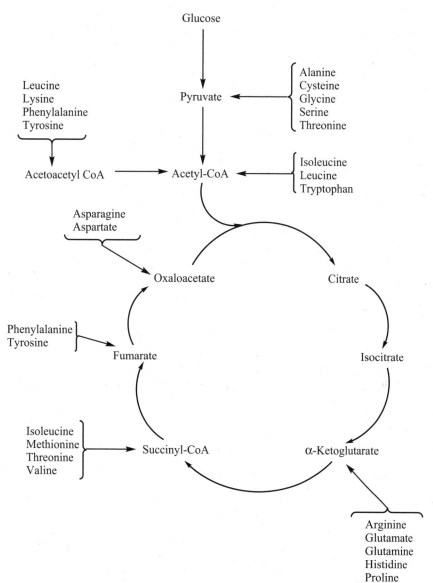

Figure 13-11:
Fate of the
amino acids.

Electron Transport and Oxidative Phosphorylation

The production of NADH and $FADH_2$ by the citric acid cycle supplies the materials for the next phase: oxidative phosphorylation. These reduced co-enzymes transport the electrons derived from the oxidation of pyruvate. The final fate of these electrons is the reduction of oxygen to water.

The details of oxidative phosphorylation aren't as easy to study as glycolysis and the citric acid cycle because the processes take place within the mito-chondria, where many of the proteins involved are integrated into the walls. In addition, many of the processes are coupled. The separate components of a *coupled* process must be in close proximity and often need to be in a spe-cific arrangement.

The electron transport system

A number of species in the mitochondria must undergo oxidation-reduction reactions. Oxidation involves a loss of electrons, whereas reduction involves a gain of electrons. These processes are coupled in that the electrons lost must equal the electrons gained. The *reduction potential* indicates how easily a mole-cule undergoes oxidation or reduction. The molecular players that are important to the electron transport system are the pyridine-linked dehydrogenases, flavin-linked dehydrogenases, iron-sulfur proteins, ubiquinones, and cytochromes.

Off on a tangent: Dealing with reduction potentials

The standard for reduction potentials is the reaction

$$2\ H^+\ (aq) + 2\ e^- \rightleftarrows H_2\ (g)$$

Under standard conditions (77 degrees Fahrenheit [25 degrees Celsius], P_{H2} = 1 atm, and $[H^+]$ = 1.0 M), the standard reduction potential is $E° = 0.00$ V. Under physiological conditions in humans, the value is -0.42 V (designated as $E'°$), because the conditions aren't standard.

Table 13-1 lists a number of physiological reduction potentials. We show you how to use these entries later. The values in the table are arranged in order of increasing potential. The higher the value, the better the reaction is at oxi-dation, and the lower the value, the better the reaction is at reduction.

Table 13-1 Some Physiological Reduction Potentials (E'°)	
Reduction Potentials	*E'°(volts)*
Ferredoxin-Fe^{3+} + e^- ⇌ Ferredoxin-Fe^{2+}	−0.43
$2 H^+(aq) + 2 e^- ⇌ H_2(g)$	−0.42
α-Ketoglutarate + CO_2 + $2 H^+$ + $2 e^-$ ⇌ Isocitrate	−0.38
$NAD^+ + H^+ + 2 e^-$ ⇌ NADH	−0.32
FAD + $2 H^+$ + $2 e^-$ ⇌ $FADH_2$	−0.22
Riboflavin + $2 H^+$ + $2 e^-$ ⇌ Riboflavin-H_2	−0.20
Dihydroxyacetone phosphate + $2 H^+$ + $2 e^-$ ⇌ Glycerol 3-phosphate	−0.19
Pyruvate + $2 H^+$ + $2 e^-$ ⇌ Lactate	−0.19
Oxaloacetate + $2 H^+$ + $2 e^-$ ⇌ L-Malate	−0.17
Fumarate + $2 H^+$ + $2 e^-$ ⇌ Succinate	+0.03
Cytochrome b-Fe^{3+} + e^- ⇌ Cytochrome b-Fe^{2+}	+0.08
Cytochrome c-Fe^{3+} + e^- ⇌ Cytochrome c-Fe^{2+}	+0.22
Cytochrome c_1-Fe^{3+} + e^- ⇌ Cytochrome c_1-Fe^{2+}	+0.23
Cytochrome a-Fe^{3+} + e^- ⇌ Cytochrome a-Fe^{2+}	+0.29
Cytochrome a_3-Fe^{3+} + e^- ⇌ Cytochrome a_3-Fe^{2+}	+0.38
½ O_2 + $2 H^+$ + $2 e^-$ ⇌ H_2O	+0.82

Each reaction in Table 13-1 is known as a *half-reaction*. Two half-reactions —
one oxidation and one reduction — are necessary to produce a complete
(oxidation-reduction) reaction. The electrons lost (oxidation) must equal the
electrons gained (reduction). For this reason, electrons appear only in the
half-reaction and never in the overall reaction.

By convention, the reactions in Table 13-1 all appear as *reduction* half-reactions.
To convert any of these to an *oxidation* half-reaction, you must do two things.
First, reverse the reaction, and then reverse the sign of E'°. In an oxidation-
reduction reaction, the overall reaction is created by combining (adding) an
oxidation reaction with a reduction reaction. Before adding the two reactions,
though, make sure that the electrons in each reaction are equal. This may
require multiplying one or both of the reactions by a value to make sure the

electrons are equal. (Multiply the reactions only — don't change the value of E'° [other than a sign change].) For example, look at the following reactions from the table:

$$\text{NAD}^+ + \text{H}^+ + 2 \text{ e}^- \rightleftarrows \text{NADH} \qquad\qquad -0.32$$

$$\text{Cytochrome b-Fe}^{3+} + \text{e}^- \rightleftarrows \text{Cytochrome b-Fe}^{2+} \qquad +0.08$$

Now change the first reaction to an oxidation:

$$\text{NADH} \rightleftarrows \text{NAD}^+ + \text{H}^+ + 2 \text{ e}^- \qquad\qquad +0.32$$

If you now want to combine these reactions, you need to multiply the cytochrome reaction by 2 (so both reactions now involve two electrons):

$$2 \text{ Cytochrome b-Fe}^{3+} + 2 \text{ e}^- \rightleftarrows 2 \text{ Cytochrome b-Fe}^{2+} \qquad +0.08$$

The number of electrons lost must equal the electrons gained. Also, notice that only the reaction is doubled, not the voltage. You can now combine these two reactions, canceling the electrons from both sides:

$$\text{NADH} \rightleftarrows \text{NAD}^+ + \text{H}^+ + 2 \text{ e}^- \qquad\qquad +0.32 \text{ V}$$

$$\underline{2 \text{ Cytochrome b-Fe}^{3+} + 2 \text{ e}^- \rightleftarrows 2 \text{ Cytochrome b-Fe}^{2+}} \qquad \underline{+0.08 \text{ V}}$$

$$\text{NADH} + 2 \text{ Cytochrome b-Fe}^{3+} \rightleftarrows 2 \text{ Cytochrome b-Fe}^{2+} + \text{NAD}^+ + 2\text{H}^+ \quad +0.40 \text{ V}$$

The final reaction has no electrons. Other species may cancel if they appear on both sides of the reaction arrow. Anytime the sum of the two potentials is positive, the reaction produces energy. Conversely, a negative value means the reaction requires energy. The greater the value of the sum, the greater the amount of energy produced.

Pyridine-linked dehydrogenases

In order for these enzymes to function, the coenzyme NAD^+ or NADP^+ is necessary. The coenzymes may be in either the oxidized or reduced forms. If the general form of the substrate in the reduced form is Z-H_2 and in the oxidized form is Z, then the reaction is as follows:

$$\text{Z-H}_2 + \text{NAD}^+ \text{ (or NADP}^+) \rightleftarrows \text{Z} + \text{NADH (or NADPH)} + \text{H}^+$$

More than 200 pyridine-linked dehydrogenases exist. Most NAD^+-linked dehydrogenases are involved in aerobic respiration. Most NADP^+-linked dehydrogenases are involved in biosynthesis.

Flavin-linked dehydrogenases

Enzymes (E) of this type require FAD or FMN as tightly bound prosthetic groups or coenzymes. Again, the species may be in either the oxidized or the reduced form. The general reactions of this type are

$$Z\text{-}H_2 + E\text{-}FAD \rightleftarrows Z + E\text{-}FADH_2$$
$$Z\text{-}H_2 + E\text{-}FMN \rightleftarrows Z + E\text{-}FMNH_2$$

NADH dehydrogenase, which contains the prosthetic group FMN, is the enzyme responsible for transporting electrons from NADH to the next acceptor in the electron transport chain. There are other flavin-linked dehydrogenases, such as succinate dehydrogenase.

Iron-sulfur proteins

The chief characteristic of iron-sulfur proteins is the presence of iron and sulfur, as S^{2-}. The electron-transporting ability of these proteins is the Fe^{2+}/Fe^{3+} couple. Several of these proteins are associated with the electron transport chain, where they're complexed to other respiratory species. Examples include succinate dehydrogenase, with two iron-sulfur centers, and NADH dehydrogenase, with four iron-sulfur centers.

Ubiquinones

The *ubiquinones* are a group of coenzymes that are fat-soluble. Coenzyme Q (CoQ) is an example of a ubiquinone. The oxidation-reduction center is a derivation of quinone, and the fat-solubility is enhanced by the presence of a long hydrocarbon chain containing a series of isoprene units. Many of the ubiquinones differ only in the number of isoprene units present. The oxidized form of coenzyme Q is simply CoQ, whereas the reduced form is $CoQH_2$. Figure 13-12 shows the general structures of both the oxidized and reduced forms of a ubiquinone.

Cytochromes

The *cytochromes* are a group of proteins that contain a heme group. Like the iron-sulfur proteins, the oxidation-reduction couple is Fe^{2+}/Fe^{3+}. The three general classes of cytochromes are a, b, and c. The derivation of the class names relates to spectral studies done during the first isolation of these molecules. Cytochromes occur in both the mitochondria and the endoplasmic reticulum. The heme group, present in all cytochromes, is like the heme groups present in myoglobin and hemoglobin. In all cases, the central portion of the group is identical; differences derive from the attachment of side chains to the heme core. Figure 13-13 shows the heme core and where the side chains normally attach.

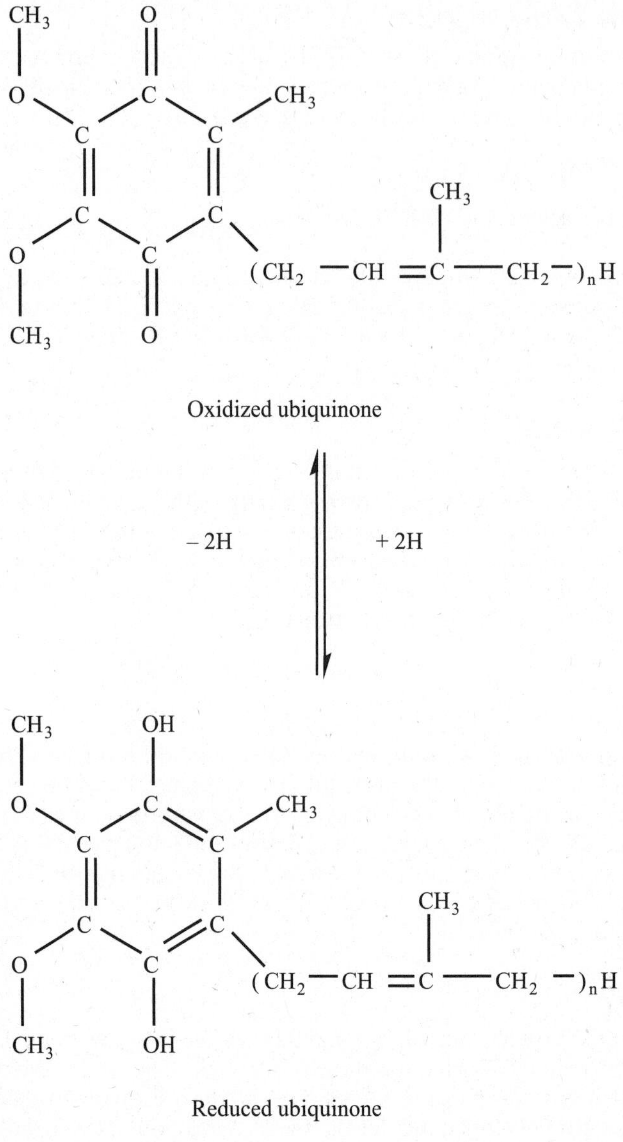

Oxidized ubiquinone

$- 2H$ $+ 2H$

Figure 13-12:
General
structures
of the oxi-
dized and
reduced
forms of a
ubiquinone.

Reduced ubiquinone

Five cytochromes (a, a_3, b, c, and c_1) have been identified as part of the electron transport chain of mammals. Cytochrome c, or *cyt c,* is easy to extract
from cells and is therefore the most studied of the cytochromes. The structure of cytochrome c from different species is important to the study of

biochemical evolution. The differences in the cytochrome c gene itself are also useful to evolutionary geneticists examining the differences among species and populations.

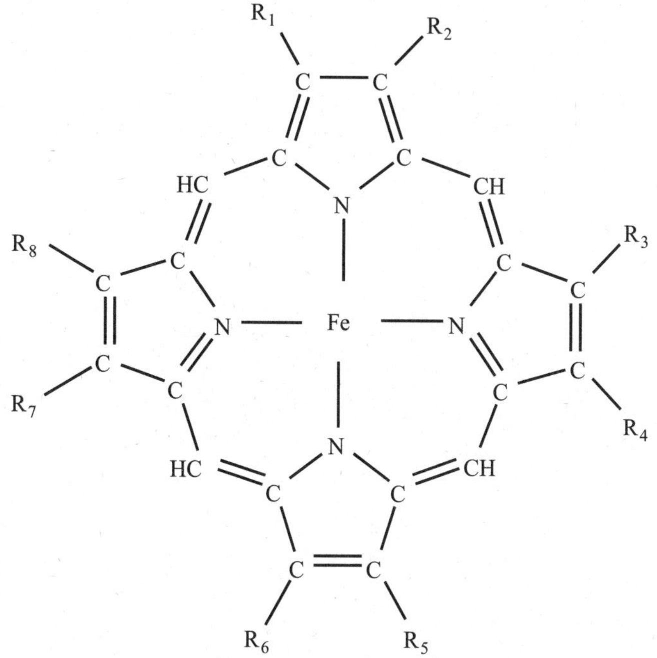

Figure 13-13:
The heme core and attachment sites (R).

Cytochromes a and a_3, *cyt a* and *cyt a_3* occur together as a complex that contains not only the expected two heme groups but also two copper ions. The copper ions are part of another oxidation-reduction couple (Cu^+/Cu^{2+}). This complex, known as *cytochrome oxidase,* is the terminal cytochrome, which transfers electrons to O_2.

Interpersonal relationships (No, it's not what you think)

The members of the electron transport chain are grouped into four complexes, with coenzyme Q (CoQ) and cytochrome c (cyt c) serving as links. One way of indicating the sequence of events in the electron transport chain appears in Figure 13-14. Figure 13-15 illustrates the same sequence emphasizing the cyclic nature of the steps. The processes take place in four complexes linking CoQ and cytochrome c. These complexes are part of the inner mitochondrial membrane.

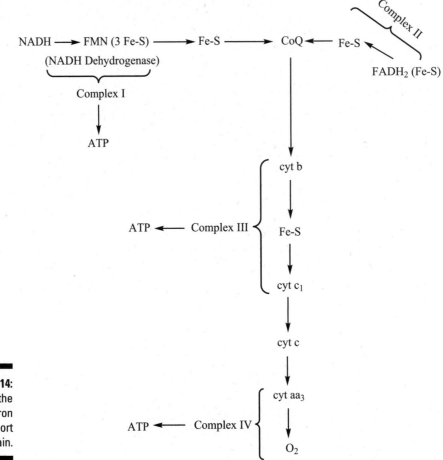

Figure 13-14:
Steps in the
electron
transport
chain.

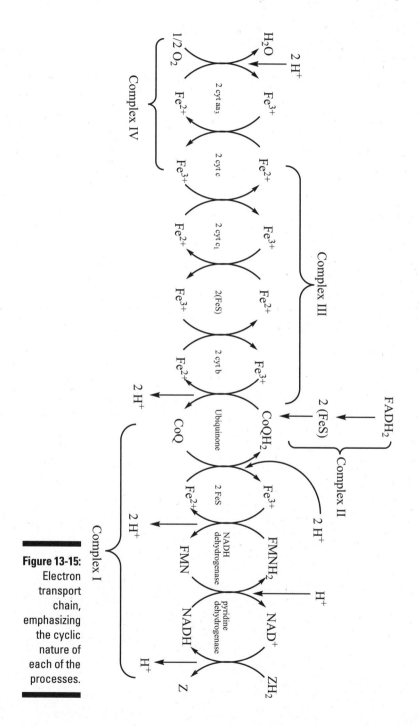

Figure 13-15:
Electron
transport
chain,
emphasizing
the cyclic
nature of
each of the
processes.

Oxidative phosphorylation

The processes of oxidative phosphorylation and the electron transport chain are closely coupled. Oxidizing the reduced forms of the coenzymes $FADH_2$ and NADH is only possible in the presence of ADP. The oxidations couple with the ADP, transforming to ATP (phosphorylation).

If you calculate the oxidation-reduction potentials for NADH and $FADH_2$ reducing oxygen, you get

$NADH \rightleftarrows NAD^+ + H^+ + 2\ e^-$	+0.32 V
$\frac{1}{2}\ O_2 + 2\ H^+ + 2\ e^- \rightleftarrows H_2O$	+0.82 V
$\frac{1}{2}\ O_2 + H^+ + NADH \rightleftarrows H_2O + NAD^+$	+1.14 V

And

$FADH_2 \rightleftarrows FAD + 2\ H^+ + 2\ e^-$	+0.22 V
$\frac{1}{2}\ O_2 + 2\ H^+ + 2\ e^- \rightleftarrows H_2O$	+0.82 V
$\frac{1}{2}\ O_2 + FADH_2 \rightleftarrows H_2O + FAD$	+1.04 V

In both cases, the combination of the potentials is positive. Positive potentials refer to spontaneous processes, and spontaneous processes produce energy. Each NADH is capable of supplying sufficient energy to produce three ATP, and each $FADH_2$ can produce two ATP.

Proposed mechanisms

The current proposed mechanism for oxidative phosphorylation is the *chemiosmotic hypothesis*. This hypothesis assumes that the hydrogen ion gradient is a significant factor promoting the conversion of ADP to ATP. The processes occurring in the four complexes present in the inner mitochondrial membrane result in a net transfer of hydrogen ions across the membrane.

The hydrogen ion transfer results in an increase in the hydrogen ion concentration in the space between the inner and outer mitochondrial membranes. Hydrogen ions must be moved back across the membrane. This transfer of hydrogen ions is necessary in the synthesis of ATP.

ATP production

The reactions from the oxidation of glucose (glycolysis) and the oxidation of pyruvate result in the production of 36 molecules of ATP from every molecule of glucose. These reactions are:

Oxidation of Glucose:

Glucose + 2 NAD$^+$ + 2 ADP + 2 P$_i$ → 2 Pyruvate + 2 NADH + 2 H$^+$ + 2 H$_2$O + 6 ATP

Oxidation of Pyruvate:

2 Pyruvate + 5 O$_2$ + 30 ADP + 30 P$_i$ → 6 CO$_2$ + 34 H$_2$O + 30 ATP

Sum:

Glucose + 2 NAD$^+$ + 5 O$_2$ + 32 ADP + 32 P$_i$ → 6 CO$_2$ + 36 H$_2$O + 2 NADH + 2 H$^+$ + 36 ATP

These equations are a summary of the ATP production. Refer to Table 12-3 in Chapter 12 for the detailed reaction sequence.

Involving the fats: β-oxidation cycle

Fatty acids may also serve as a source of ATP (so that's what my belly is good for — energy storage). Accomplishing this requires a series of reactions, known as β-*oxidation,* or the *fatty acid spiral,* to break down the fatty acid molecule. This series of reactions is a cyclic process. Some of the processes are oxidations, which require the coenzymes NAD$^+$ and FAD. This process also occurs in the mitochondria. The initiation of fatty acid oxidation requires activation of the relatively unreactive fatty acid molecule. The activated form is analogous to acetyl-CoA. In this case, the coenzyme A binds to the fatty acid to form a fatty acyl-CoA. Activation requires acyl-CoA synthetase and one molecule of ATP. The ATP uses two phosphates and becomes AMP.

At the inner mitochondrial membrane, the fatty acyl-CoA combines with the carrier molecule carnitine. Carnitine acyltransferase catalyzes this transfer. The fatty acyl-carnitine transports into the mitochondrial matrix, where it converts back to fatty acyl-CoA. With the mitochondrial matrix, a series of four reactions constitute the cycle known as β-oxidation. The name of this process refers to the oxidation of the second (β) carbon followed by the loss of two carbons from the carboxyl end of the fatty acyl-CoA. Each trip around the cycle results in the removal of two carbon atoms, and the cycle continues until all the carbon atoms are removed. Figure 13-16 illustrates the general aspects of the cycle.

The first step in the cycle is an oxidation, with the catalyst being acyl-CoA dehydrogenase. During this step, coenzyme FAD accepts two hydrogen atoms. One of the hydrogen atoms is from the α carbon, and the other is from the β carbon atom. The process is stereospecific, producing the trans form. Elsewhere, the FADH$_2$ undergoes re-oxidation to FAD with the production of 1.5 molecules of ATP.

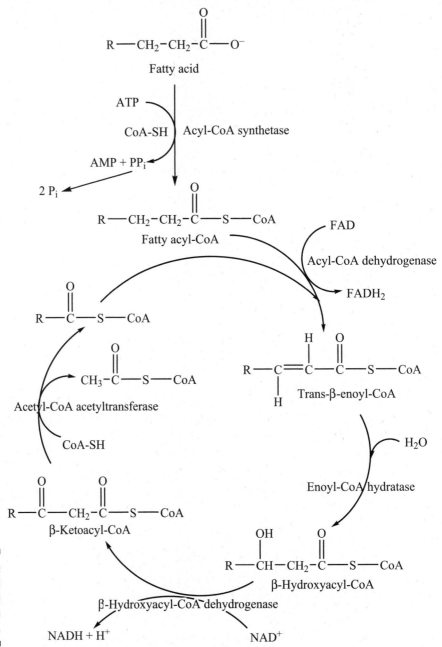

Figure 13-16:
General
steps in the
β-oxidation
cycle.

The trans-alkene undergoes hydration to form a secondary alcohol in the second step. The catalyst is the enzyme enoyl-CoA hydratase — a stereospecific enzyme yielding only the L isomer. Next, the secondary alcohol undergoes oxidation to form a ketone. The oxidizing agent is NAD^+. The enzyme catalyzing this oxidation is β-hydroxy-acyl-CoA dehydrogenase. The re-oxidation of NADH to NAD^+ via the electron transport chain produces two molecules of ATP.

The final step involves the cleavage (no, not that kind!) of the β-ketoacyl-CoA with a molecule of CoA. This produces acetyl-CoA and a fatty acyl-CoA two carbon atoms shorter than the original. The enzyme from this step is β-ketothiolase (or simply thiolase). The new fatty acyl-CoA goes around the cycle to be shortened by two carbon atoms. An unsaturated fatty acid also goes through similar steps but needs one or two additional enzymes.

The energy yield from a fatty acid is larger than from glucose. The process begins with the activation of the fatty acid, which costs the equivalent of two ATP molecules. Each trip around the cycle yields ten molecules of ATP, a molecule of $FADH_2$, and a molecule of NADH. The NADH and $FADH_2$ ultimately yield four additional molecules of ATP. Thus, each trip around the cycle produces fourteen molecules of ATP. In addition, the final trip around the cycle produces not one but two molecules of acetyl-CoA.

Not so heavenly bodies: Ketone bodies

Some of the excess acetyl-CoA forms a group of relatively small molecules called *ketone bodies.* This is especially important when there's a buildup of acetyl-CoA. A buildup may occur when the rate of production is too high or if acetyl-CoA is not used efficiently. Two acetyl-CoA molecules combine in the reverse of the last step in β-oxidation to produce acetoacetyl-CoA. Acetoacetyl-CoA reacts with water and another acetyl-CoA to form β-hydroxy-β-methylglutaryl-CoA, which in turn cleaves to acetoacetate and acetyl-CoA. Most of the acetoacetate undergoes reduction to β-hydroxybutyrate (a small amount decarboxylates to acetone and carbon dioxide). These steps appear in Figure 13-17.

The other guy

When triglyceride breaks down, the results are a glycerol and three fatty acid molecules. The fatty acid molecules enter the β-oxidation cycle and produce energy. Catabolism of the glycerol also serves as a source of energy. First, the glycerol is phosphorylated to glycerol 1-phosphate (= glycerol 3-phosphate). This uses one molecule of ATP. Oxidation of glycerol 1-phosphate generates dihydroxyacetone phosphate, which can enter the glycolysis pathway. The net energy production is from 16.5 to 18.5 molecules of ATP.

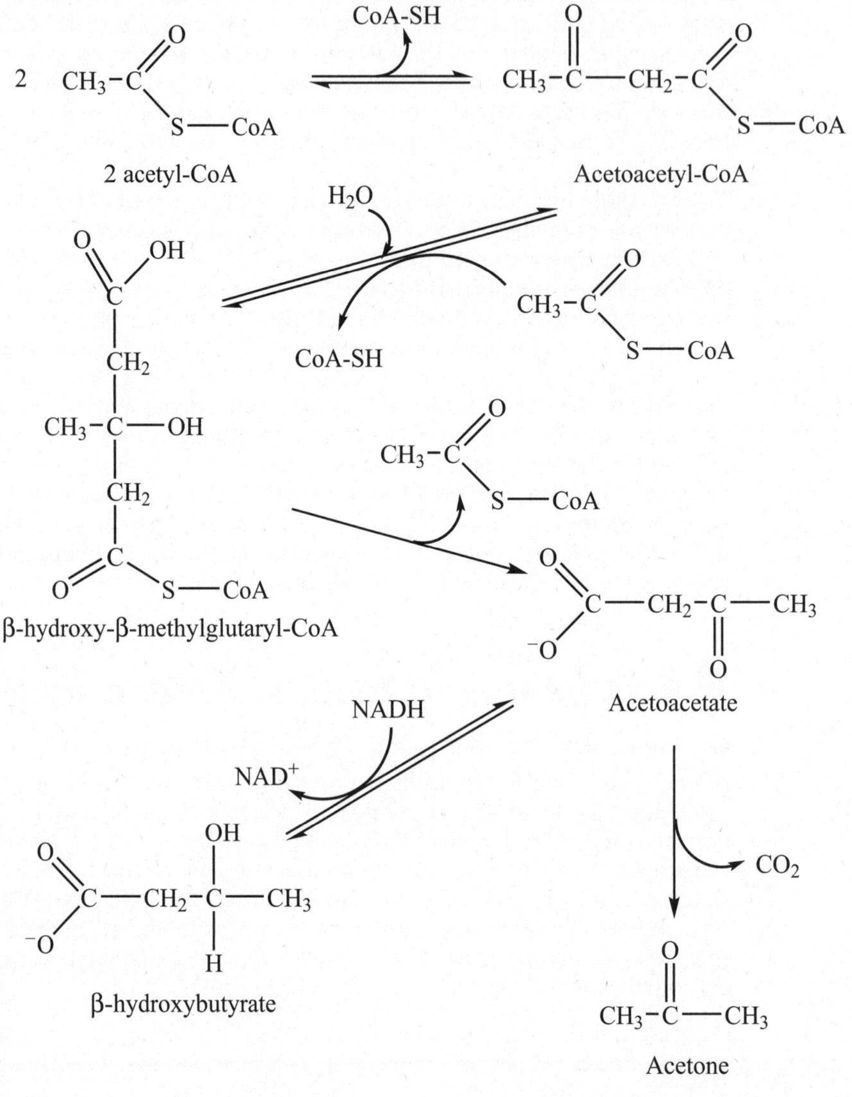

Figure 13-17:
Formation of
the ketone
bodies.

As a group, acetone, β-hydroxybutyrate, and acetoacetate are the ketone
bodies.

Ketone body formation occurs primarily in the liver, and the β-hydroxybutyrate and acetoacetate then enter the bloodstream for use by other tissues. During prolonged starvation, ketone bodies may serve as the major energy source for some tissues. The kidneys excrete excess ketone bodies. Normal blood levels are about 1 milligram of ketone bodies per 100 milliliters of blood.

In starvation or diabetes mellitus, a form of diabetes, cells may not receive sufficient carbohydrate for energy, leading to an increase in the rate of fatty acid oxidation to compensate for the energy deficit. As the amount of acetyl-CoA increases, not enough oxaloacetate is available in the citric acid cycle for oxidation of this acetyl-CoA. (The oxaloacetate concentration is lower because of the necessity of using it for glucose synthesis.) This leads to an increase in the production of ketone bodies and an increase of ketone bodies in the bloodstream. At 3 milligrams of ketone bodies per 100 milliliters, a condition known as *ketonemia* arises — a high concentration of ketone bodies in the urine. Ketonemia and ketonuria are two aspects of ketosis.

Two of the ketone bodies are in the form of acids. The buildup of ketone bodies leads to an overwhelming of the blood buffers. The decrease in blood pH may reach 0.5 units lower than the normal pH (7.4), leading to *acidosis,* a serious condition that, among other things, leads to difficulty in oxygen transport by hemoglobin. Dehydration results as the kidneys eliminate large quantities of liquid in an effort to remove the excess acid. Severe acidosis may result in a coma that may result in death.

Mammals can't convert acetyl-CoA to carbohydrates. It's possible to convert carbohydrates to fats but not to do the reverse.

Investing in the Future: Biosynthesis

One aspect of metabolism — catabolism — is to produce the energy required for life. Another aspect — anabolism — is to supply the materials for growth and replacement. Food supplies the raw fuel for metabolism. A number of pathways are available to allow for flexibility. Some pathways must be blocked to overcome Le Châtelier's Principle, partly because an enzyme could catalyze both the forward and reverse reactions.

Nearly all intermediates in catabolic processes are also intermediates in anabolic processes. Molecules may easily change from one pathway to another. In general, anabolic processes require the energy produced by catabolic processes. You've already read about one aspect of anabolism — gluconeogenesis. In the earlier section "Going in reverse: Gluconeogenesis," you saw how this process, related to glycolysis, can generate glucose and other carbohydrates. We examine other biosynthesis processes in this section.

Fatty acids

Production of the fatty acids is necessary to form the membrane lipids. But the main reason for fatty acid synthesis is to convert excess dietary carbohydrate to fats for storage (or so my thighs tell me). The key molecule for this process is acetyl-CoA.

The liver is the primary fatty acid synthesis site in humans, and humans can synthesize all the fatty acids but two: linoleic acid and linolenic acid. Linoleic acid and linolenic acid are also essential fatty acids, required components of the diet. Acetyl-CoA from glycolysis or β-oxidation reacts with the bicarbonate ion in a reaction powered by ATP and catalyzed by acetyl-CoA carboxylase, forming the three-carbon molecule malonyl-CoA (see Figure 13-18).

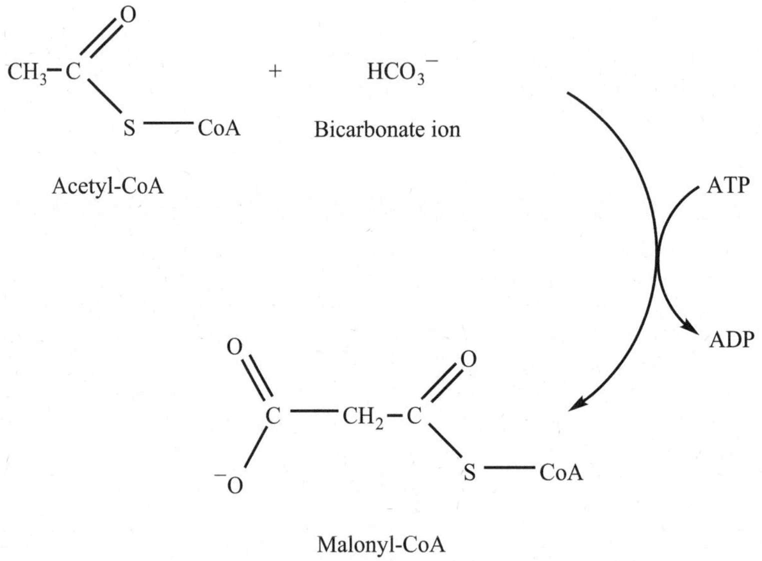

Figure 13-18:
Synthesis
of malonyl-
CoA.

The release of insulin triggers a series of steps that result in the activation of acetyl-CoA carboxylase. Release of insulin indicates high food levels. Glucagon and epinephrine, both hormones, inhibit the enzyme through a series of steps. In mammals, the enzymes necessary to synthesize palmitic acid from acetyl-CoA and malonyl-CoA are present in a complex known as *fatty acid synthase.* In plants and bacteria, the enzymes are present as separate molecules. Synthesis precedes two carbon atoms at a time, which is why all the natural fatty acids contain an even number of carbon atoms.

Synthesis begins when a molecule of acetyl-CoA links to an acyl carrier protein, ACP, and a malonyl-CoA does the same with another ACP. The two ACP-linked molecules then join and release a carbon dioxide molecule, an ACP, and an acetoacetyl-ACP. Next are three steps that are the reverse of the first three steps of β-oxidation. First, NADPH reduces the ketone group to an alcohol. Then dehydration of the alcohol leaves a double bond between the second and third carbon atoms. The coenzyme NADPH again serves as a reducing agent to produce butyryl-ACP. The sequence repeats with butyryl-ACP replacing the acetyl-ACP. Figure 13-19 illustrates these steps.

The series of synthesis steps continues up to palmitic acid (16 carbon atoms). The overall reaction is

$$8 \text{ acetyl-CoA} + 14 \text{ H}^+ + 14 \text{ NADPH} + 7 \text{ ATP} \rightarrow \text{palmitic acid} + 8 \text{ CoA} + 14 \text{ NADP}^+ + 7 \text{ ADP} + 7 \text{ P}_i$$

After the palmitic acid forms, additional reactions, where necessary, can lengthen or shorten the chain. These require different enzyme systems. Partial oxidation of a saturated fatty acid yields an unsaturated fatty acid.

Membrane lipids

Like other molecules, membrane lipids are synthesized from their constituents. In the preceding section, we explain how to synthesize the fatty acids. These fatty acids need to be activated with acetyl-CoA in order to produce the appropriate acyl-CoA. The reduction of dihydroxyacetone from glycolysis yields glycerol 3-phosphate, which combines with the appropriate acyl-CoA molecules to yield a phosphatidate (see Figure 13-20). The phosphatidate then reacts with an activated serine or an activated choline to form the appropriate phosphoglyceride.

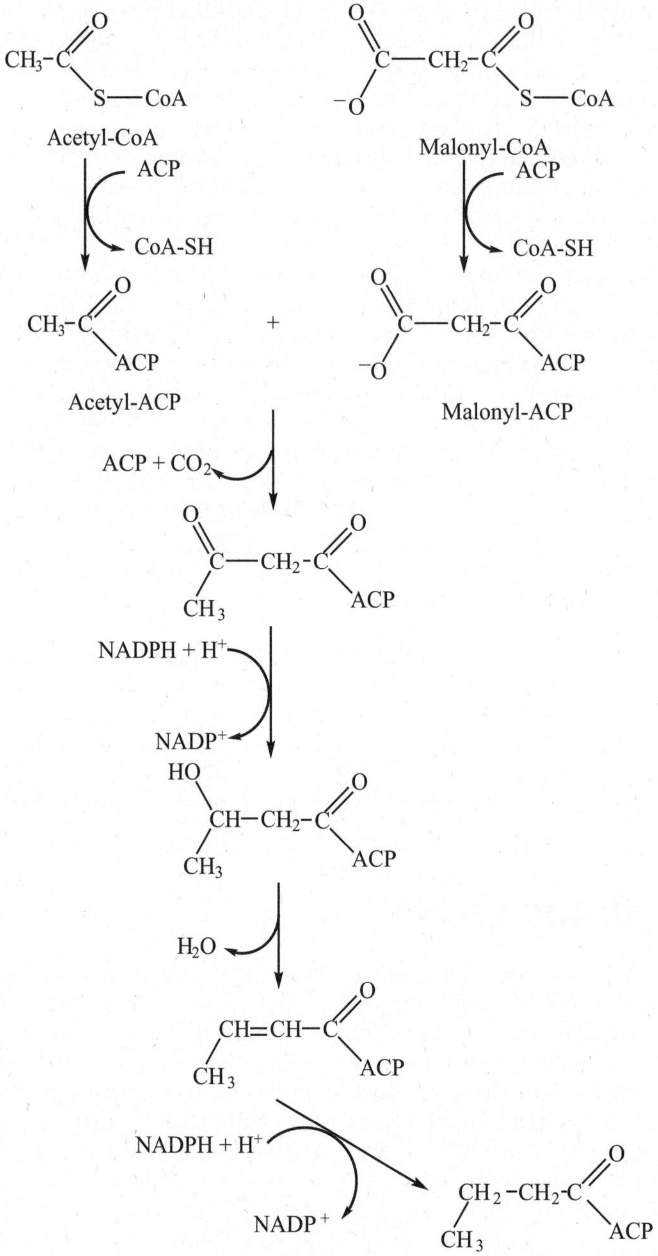

Figure 13-19:
Fatty acid
synthesis.

The formation of the spingolipids follows a similar path. In this case, sphingosine replaces glycerol. The synthesis of sphingosine begins with the reaction
of palmitoyl-CoA, with serine in the presence of acid. This reaction yields
Coenzyme A, carbon dioxide, and the precursor of sphingosine. Oxidation of
the precursor yields sphingosine (see Figure 13-21).

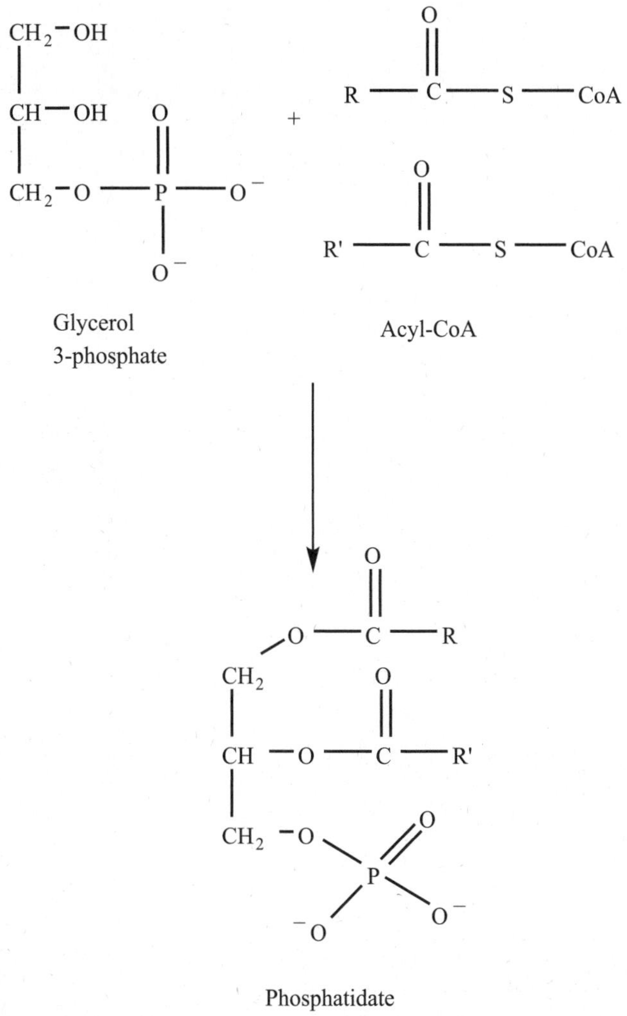

Figure 13-20:
Formation of a phosphatidate.

An acyl-CoA can then add a fatty acid to the amine group to produce N-acylsphingosine (ceramide). The reaction of the alcohol on the third carbon of the ceramide with activated phosphatidylcholine yields sphingomyelin.

The reaction of ceramide with an activated monosaccharide begins the synthesis of the glycolipids. Adding additional activated monosaccharides (UDP-glucose being one example) is necessary to complete the synthesis.

Cholesterol is another membrane lipid. It helps control the fluidity of cell membranes and is also the precursor of the steroid hormones. The entire synthesis takes place in the liver, where acetyl-CoA molecules are joined. Thus, the cholesterol molecule is built up two carbon atoms at a time.

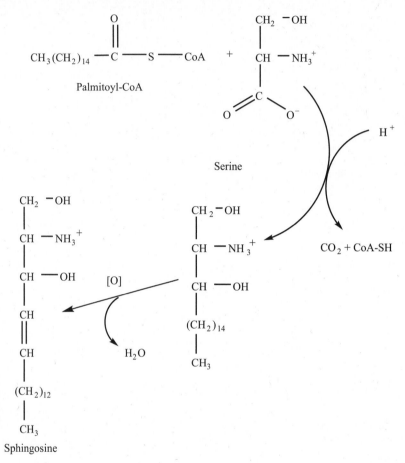

Figure 13-21:
Formation of
sphingosine.

Amino acids

Synthesis of amino acids becomes necessary when insufficient quantities
are present in the diet. Adult humans can only synthesize 11 of the 20 amino
acids. The amino acids that humans can't synthesize are known as the *essential amino acids,* and these are a necessary requirement in the diet. Table 13-2
lists the essential and nonessential amino acids.

Table 13-2	Essential and Nonessential Amino Acids
Essential Amino Acids	*Nonessential Amino Acids*
Histidine	Alanine
Isoleucine	Asparagine

Essential Amino Acids	Nonessential Amino Acids
Leucine	Aspartate
Lysine	Cysteine
Methionine	Glutamate
Phenylalanine	Glutamine
Threonine	Glycine
Tryptophan	Proline
Valine	Serine

You probably noticed that Table 13-2 lists only 18 amino acids, so what about the other two? Well, arginine is essential for children but not for adults. Tyrosine is nonessential in the presence of adequate quantities of phenylalanine. Glutamate is important to the synthesis of five amino acids. Glutamate may form by the reduction of α-ketoglutaric acid, an intermediate from the Krebs cycle. The process is shown in Figure 13-22.

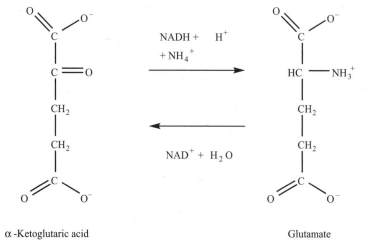

Figure 13-22:
Equilibrium between glutamate and α-ketoglutaric acid.

$NADH + H^+$
$+ NH_4^+$

$NAD^+ + H_2O$

α-Ketoglutaric acid

Glutamate

In the forward direction, this is a synthesis reaction, whereas the reverse reaction is an important oxidative deamination from the catabolism of amino acids. Glutamate, when necessary, serves as an intermediate in the biosynthesis of alanine, asparagine, aspartate, glutamine, proline, and serine. The transamination in Figure 13-23 illustrates the formation of alanine.

Replacing pyruvate in the preceding reaction with oxaloacetate yields aspartate.

Excess phenylalanine can be converted to tyrosine by a simple oxidation in the presence of phenylalanine hydroxylase (see Figure 13-24).

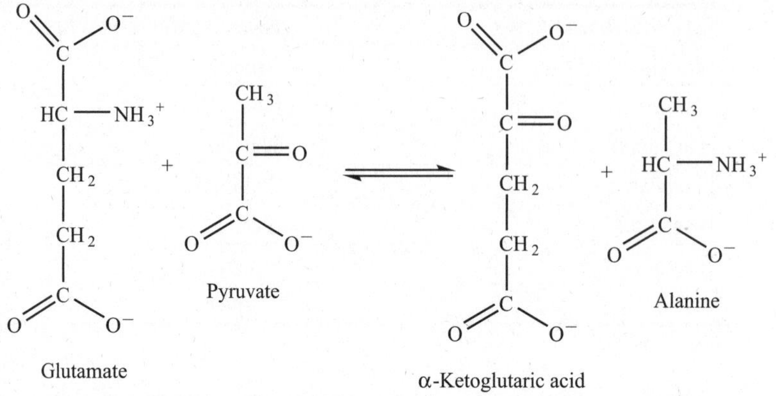

Figure 13-23:
Formation of
alanine.

Methionine serves as the source of sulfur for the synthesis of cysteine. Serine
serves as the base of the rest of the molecule. Serine is the product of a
three-step process beginning with 3-phosphoglycerate. The process starts
with the oxidation by NAD⁺ of the secondary alcohol group. The ketone thus
formed undergoes transamination with glutamate to form 3-phosphoserine.
Finally, hydrolysis of the phosphate ester yields serine (see Figure 13-25).

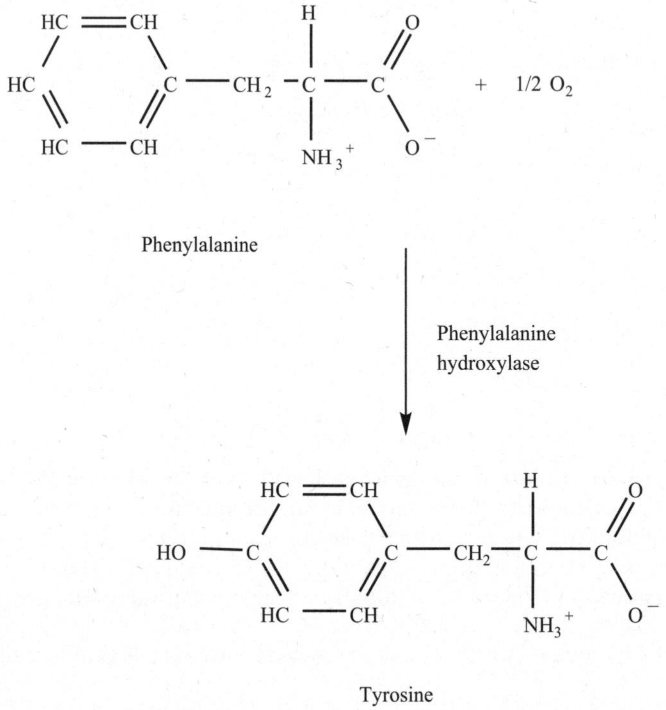

Figure 13-24:
Synthesis of
tyrosine.

The formation of proline is a four-step process beginning with glutamate. The process is shown in Figure 13-26.

The formation of proline is a four-step process beginning with glutamate. The process is shown in Figure 13-26.

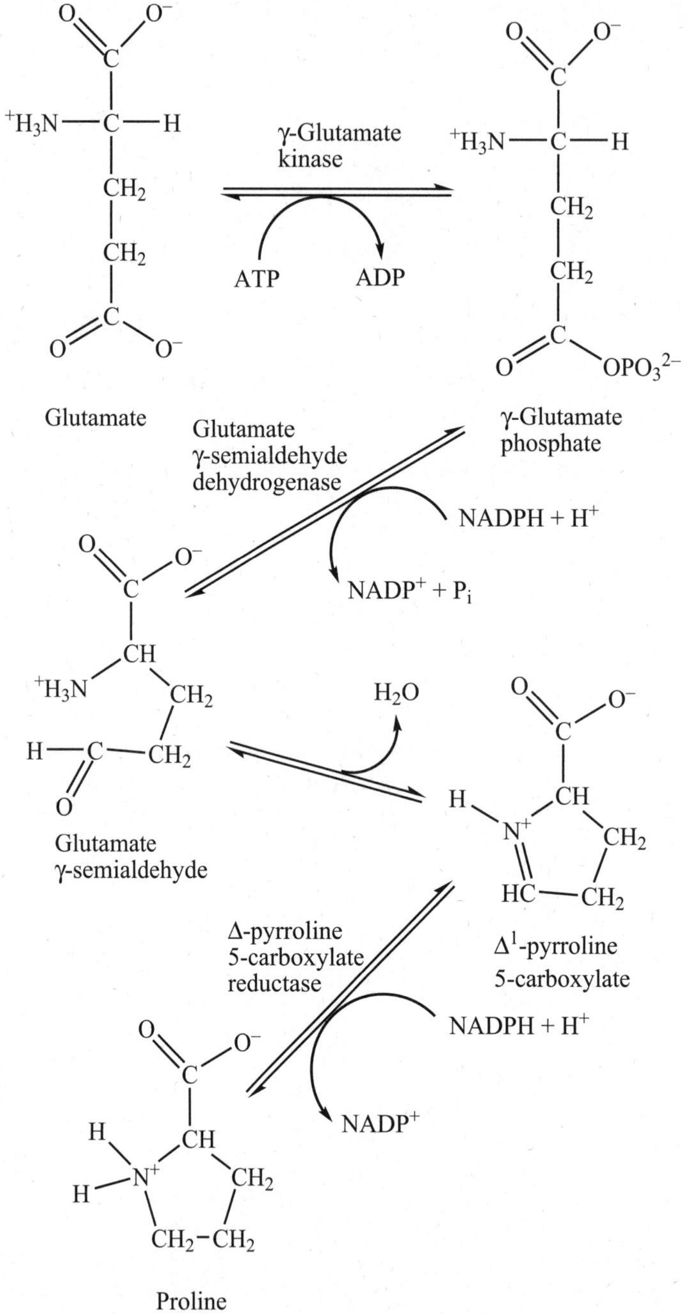

Figure 13-26:
Synthesis of
proline.

Chapter 14

Smelly Biochemistry: Nitrogen in Biological Systems

· ·

In This Chapter

▶ Talking about purine and pyrimidine synthesis

▶ Examining catabolism and discussing the urea cycle

▶ Considering amino acids

▶ Rounding up some metabolic disorders

· ·

*I*n this chapter, we investigate the role of nitrogen in biomolecules. Nitrogen occurs primarily in amino acids (proteins) and in nucleic acids (purines and pyrimidines), many of which have a distinctive and generally unpleasant aroma, hence our chapter title. A few other molecules, such as hemoglobin, also contain nitrogen. Humans eliminate nitrogen primarily as urea.

Ring in the Nitrogen: Purine

Adenine and *guanine* are nitrogen bases that employ the *purine ring system* (see Figure 14-1). The formation of these molecules is essential to the synthesis of both DNA and RNA. The biosynthesis of the purines generates the molecules in their nucleotide forms instead of the free base form.

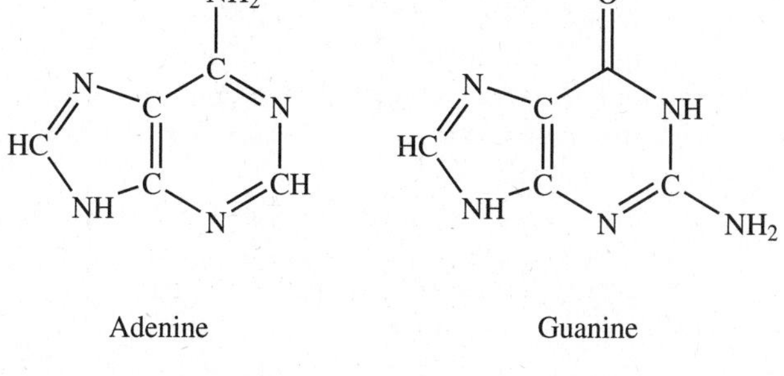

Figure 14-1: Purine nitrogen bases.

Adenine   Guanine

Biosynthesis of purine

The synthesis of purine (that's *purine,* not purée) begins with the activation of D-ribose-5'-phosphate through *pyrophosphorylation.* In this reaction a pyrophosphate group from ATP is transferred to C-1 of an α-D-ribose-5'-phosphate. This gives a 5-phospho-α-D-ribose 1-pyrophosphate (PRPP) and AMP. The reaction is unusual because it involves the transfer of an intact pyrophosphate group (see Figure 14-2). PRPP is also necessary for the synthesis of pyrimidines (see the "Pyrimidine Synthesis" section later in the chapter).

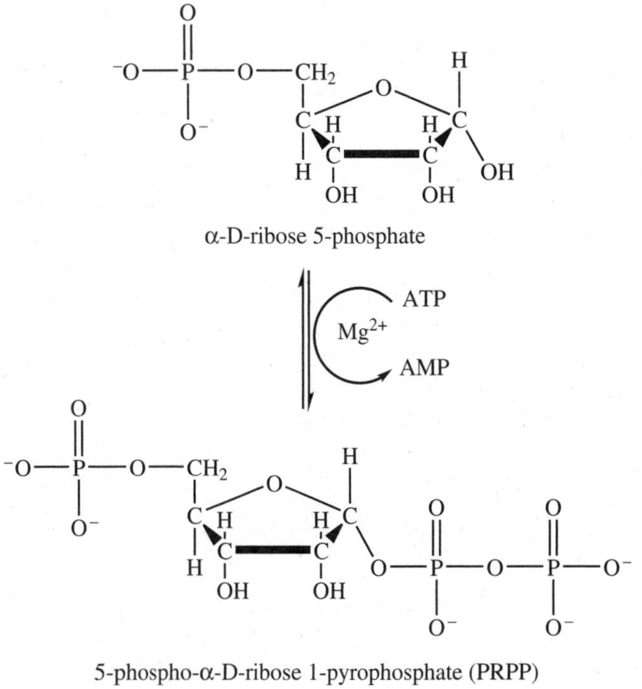

α-D-ribose 5-phosphate

ATP

Mg^{2+}

AMP

Figure 14-2:
Activation of
D-ribose-5'-
phosphate.

5-phospho-α-D-ribose 1-pyrophosphate (PRPP)

Inosine synthesis

PRPP goes through a series of ten steps (see Figure 14-3) to become inosine 5'-phosphate, or *inosinic acid* (IMP). Notice that throughout these ten steps the D-ribose-5'-phosphate portion of PRPP doesn't change. Table 14-1 shows the ten enzymes that are necessary for these steps. Two additional, though different, steps are necessary to convert IMP to either AMP or GMP.

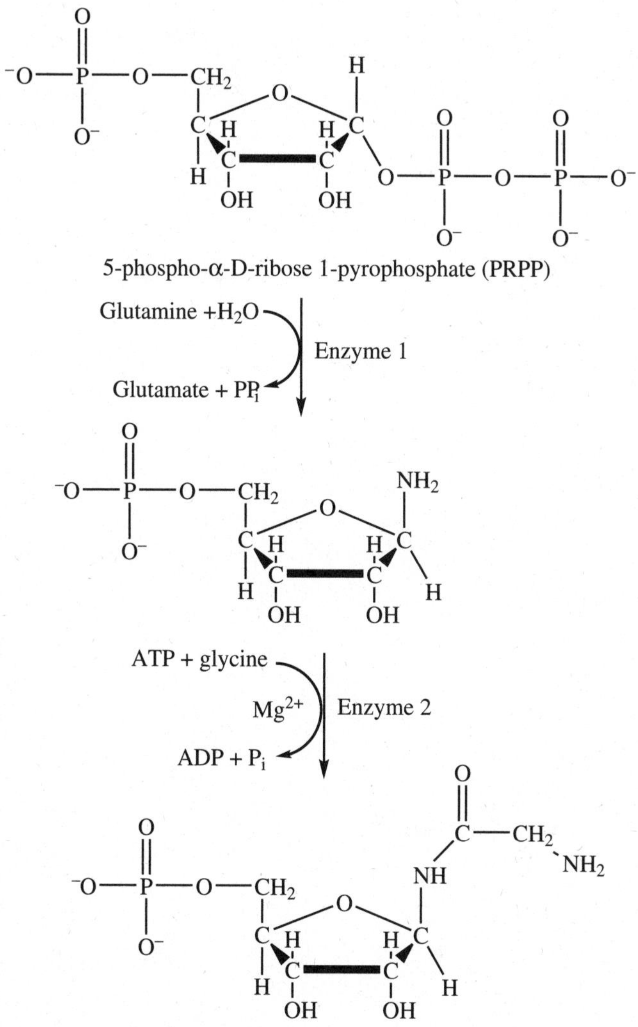

5-phospho-α-D-ribose 1-pyrophosphate (PRPP)

Glutamine +H_2O

Enzyme 1

Glutamate + PP_i

ATP + glycine

Mg^{2+} Enzyme 2

ADP + P_i

Figure 14-3:
The ten steps necessary to convert PRPP (5-phospho-α-D-ribose 1-pyro-phosphate) into inosine 5'-phos-phate.

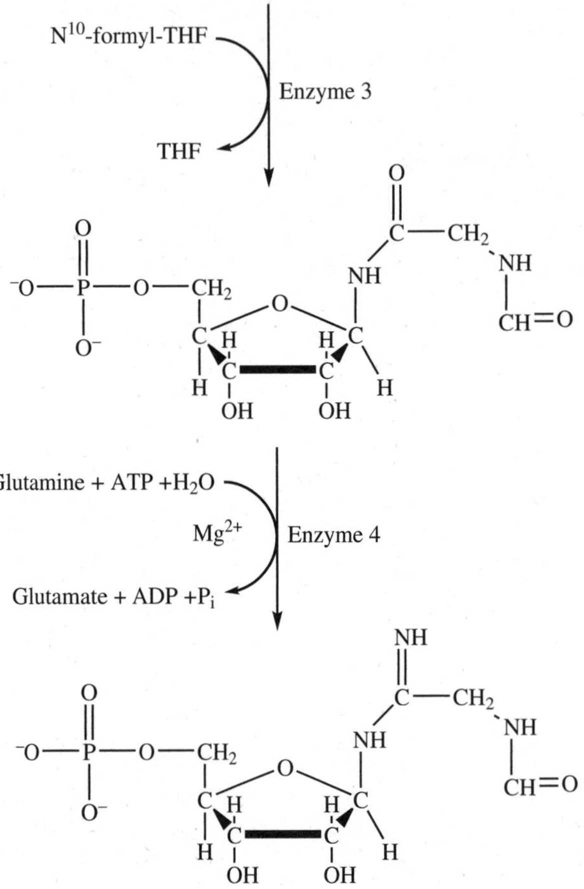

Figure 14-3:
(continued)

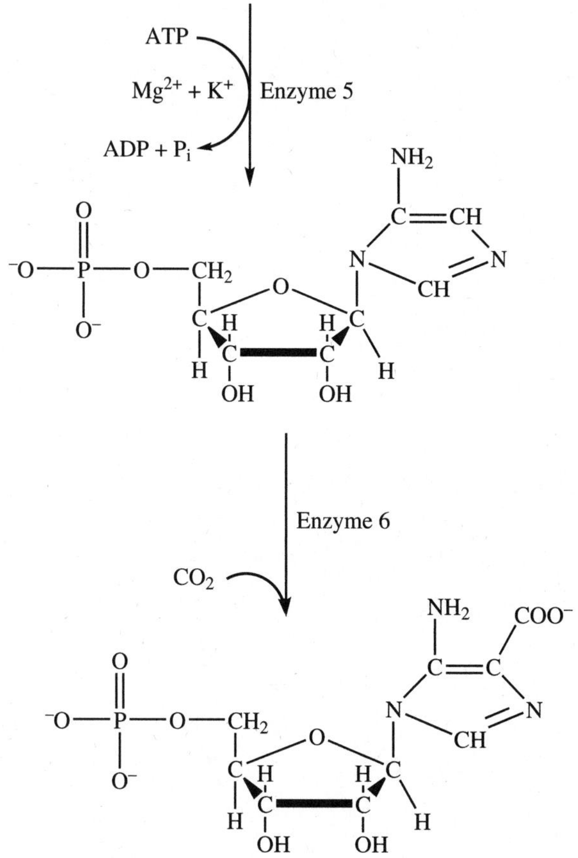

Figure 14-3:
(continued)

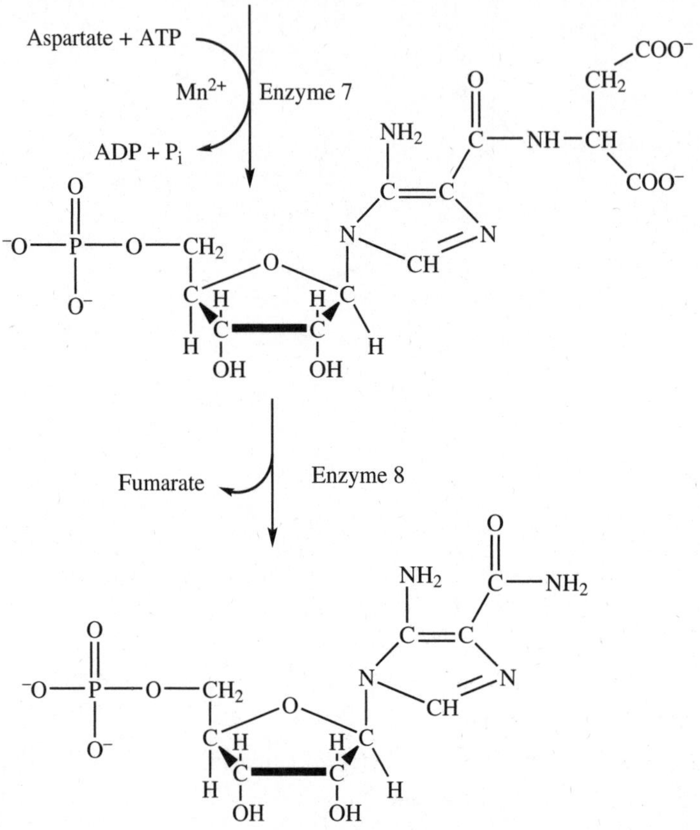

Figure 14-3: (continued)

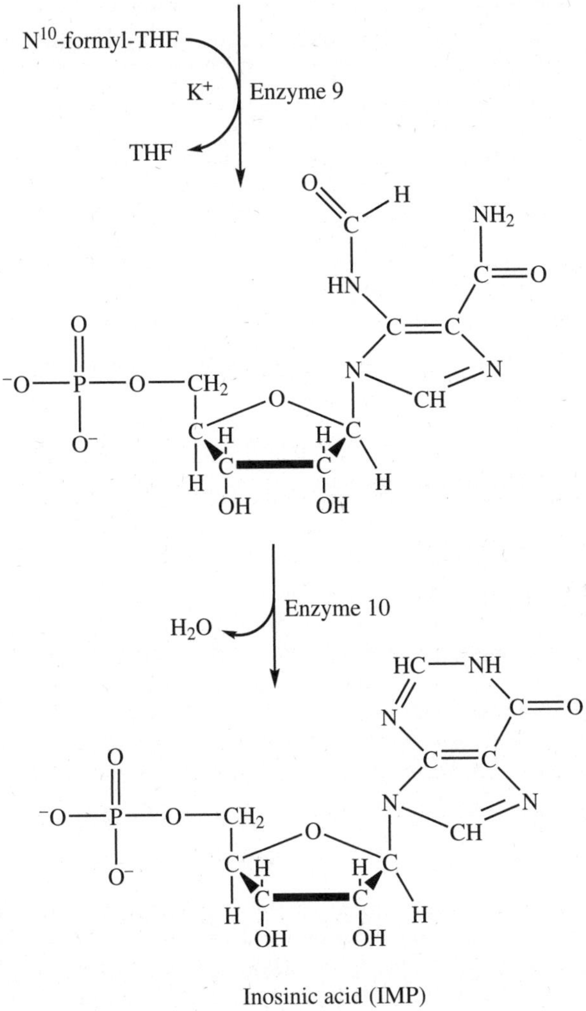

Figure 14-3:
(continued)

Inosinic acid (IMP)

Table 14-1   Ten Enzymes Necessary for Inosine Synthesis

Enzyme	Name
1	Amidophosphoribosyl transferase
2	Phosphoribosylglycinamide synthetase
3	Phosphoribosylglycinamide formyltransferase
4	Phosphoribosylformylglycinamide synthetase
5	Phosphoribosylaminoimidazole synthetase
6	Phosphoribosylaminoimidazole carboxylase
7	Phosphoribosylaminoimidazole-succinocarboxamide synthetase
8	Adenylosuccinate lyase
9	Phosphoribosylaminoimadazolecarboxamide formyltransferase
10	IMP cyclohydrolase

AMP synthesis

To convert IMP into AMP (adenosine monophosphate), it's necessary to transfer an amino group from an aspartate. This transfer requires two steps, and the energy to add aspartate to IMP comes from the hydrolysis of a GTP (guanosine triphosphate). The process is then completed by the loss of fumarate. The enzyme *adenylosuccinate synthetase* catalyzes the first step, and the enzyme *adenylosuccinate lyase* catalyzes the second. Figure 14-4 illustrates the process.

GMP synthesis

The conversion of IMP to GMP (guanosine monophosphate) begins with the IMP dehydrogenase catalyzed oxidation to xanthosine 5'-phosphate. The coenzyme for this step is NAD^+. GMP synthetase catalyzes the next step, the amine transfer from glutamate. The hydrolysis of ATP supplies the energy for this step (see Figure 14-5).

How much will it cost?

The biosynthesis of both AMP and GMP requires the hydrolysis of several high-energy bonds. To produce IMP from D-ribose 5-phosphate requires the hydrolysis of six high-energy bonds (one PP_i and five ATP). To convert IMP to AMP requires the hydrolysis of one more high-energy bond (from GTP). And to convert IMP to GMP requires the hydrolysis of two high-energy bonds — one ATP and one PP_i.

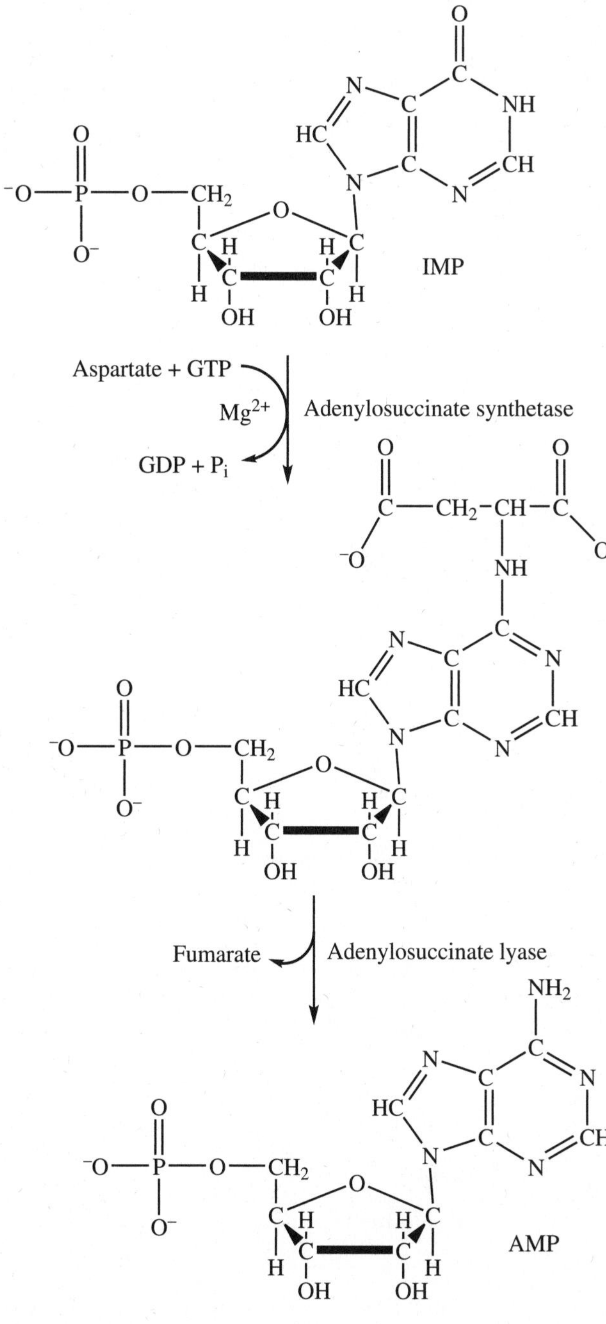

Figure 14-4:
Conversion
of IMP to
AMP.

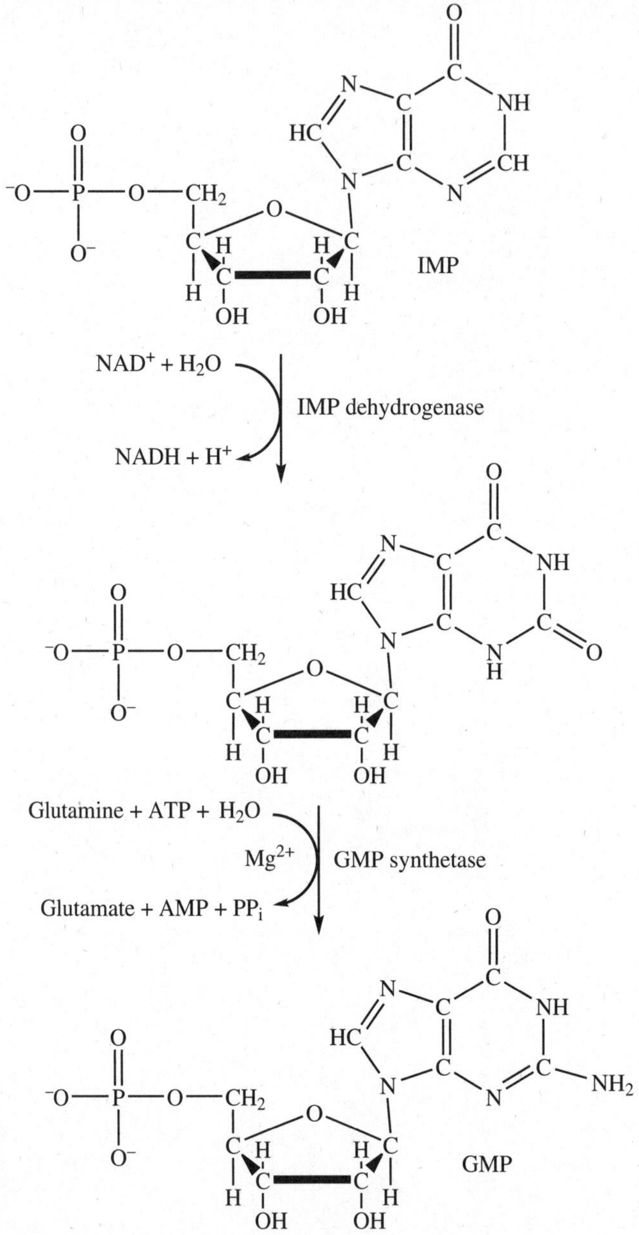

Figure 14-5:
Conversion
of IMP to
GMP.

Anaerobic organisms, such as the bacteria responsible for tetanus or botulism, must oxidize four glucose molecules at two ATP per glucose to meet the energy requirement. An aerobic organism, like you, for example, needs to oxidize only one glucose molecule at 36 ATP per glucose. These processes require a substantial amount of energy. Sometimes, metabolic processes known as the *salvage pathways* may lessen the energy requirement. In the salvage pathways, nitrogen bases are recycled instead of synthesized and are then converted to nucleotides.

Pyrimidine Synthesis

The biosynthesis of *pyrimidines* follows a different path from purine synthesis, which I cover in the preceding section. In this case, synthesis of the base takes place before attachment to the ribose. Ring synthesis requires bicarbonate ion, aspartic acid, and ammonia. Although using ammonia directly is possible, it usually comes from the hydrolysis of the side chain of glutamine.

First step: Carbamoyl phosphate

The initial step in pyrimidine synthesis is to transfer a phosphate from an ATP to a bicarbonate ion to form *carboxyphosphate,* which in turn undergoes an exchange where ammonia replaces the phosphate to form *carbamic acid.* Whew! — got that? A second ATP transfers a phosphate to carbamic acid to form *carbamoyl phosphate.* Figure 14-6 summarizes these steps.

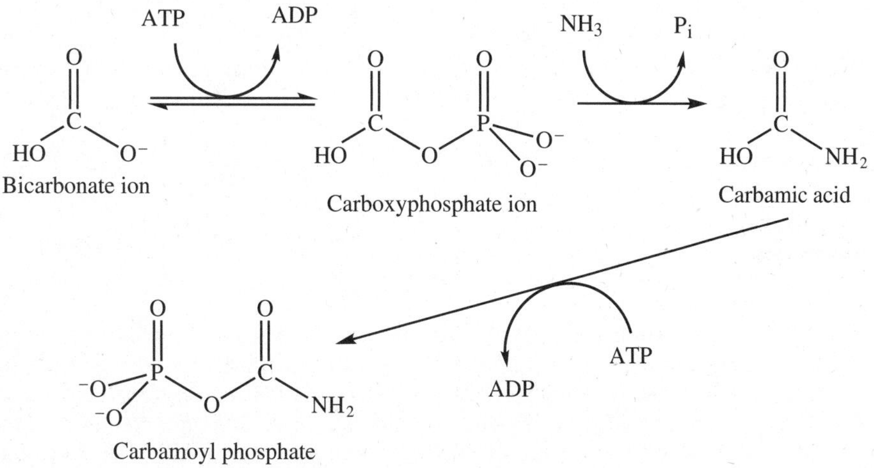

Figure 14-6:
Synthesis of carbamoyl phosphate.

The primary enzyme for the process in Figure 14-6 is *carbamoyl synthetase.* One region of the enzyme is responsible for the synthesis of carbamic acid, and a second region hydrolyzes ammonia from glutamine. A third region completes the process, and a channel connects the three regions.

Next step: Orotate

The next step in pyrimidine synthesis is the formation of *orotate,* which is joined to a ribose. It begins with the enzyme *aspartate transcarbamoylase,* which joins aspartate to carbamoyl phosphate with the loss of phosphate. This forms *carbamoylaspartate,* which cyclizes to *dihydroorotate,* which is oxidized by NAD^+ to orotate (see Figure 14-7).

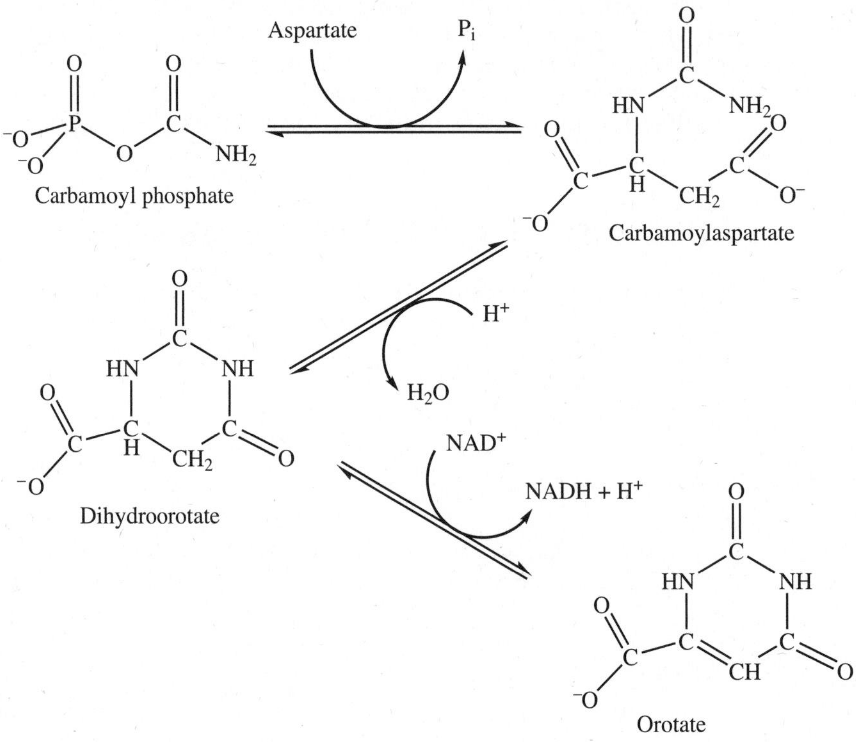

Figure 14-7: Formation of orotate from carbamoyl phosphate.

Orotate joins with PRPP (see the "Biosynthesis of purine" section earlier in the chapter) to form *orotidylate,* with pyrophosphate hydrolysis providing the necessary energy. The enzyme *pyrimidine phosphoribosyltransferase* is responsible for this reaction. The enzyme *orotidylate decarboxylase* catalyzes the decarboxylation of orotidylate to *uridylate* (UMP). Figure 14-8 illustrates these steps.

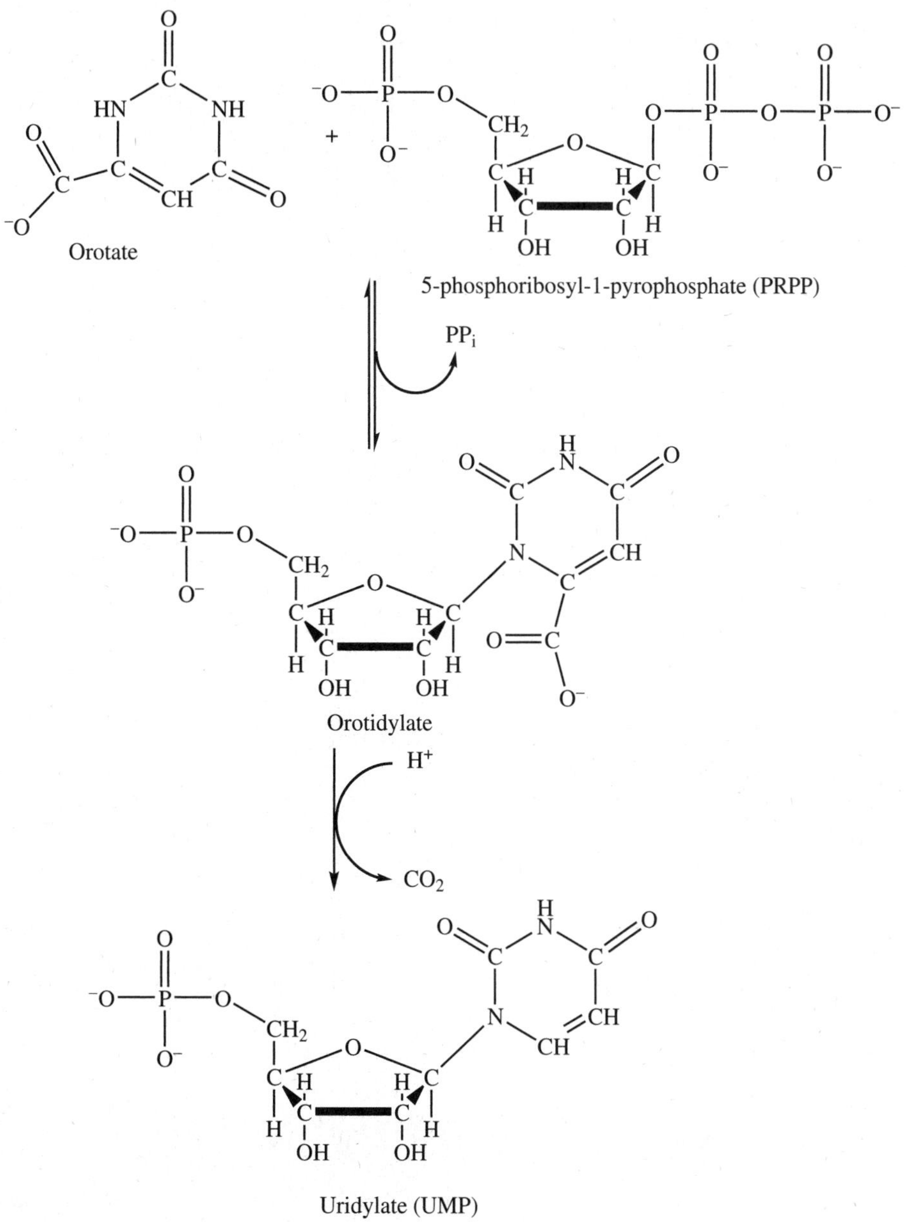

Figure 14-8:
Conversion
of orotate
to uridylate
(UMP).

Last step: Cytidine

The final nucleotide, *cytidine* (CTP), forms from uridine monophosphate (UMP). The first step is to change UMP into UTP (uridine triphosphate). UMP kinase and a nucleoside diphosphate kinase convert UMP to UTP. Figure 14-9 shows the final step — the conversion of UTP to CTP.

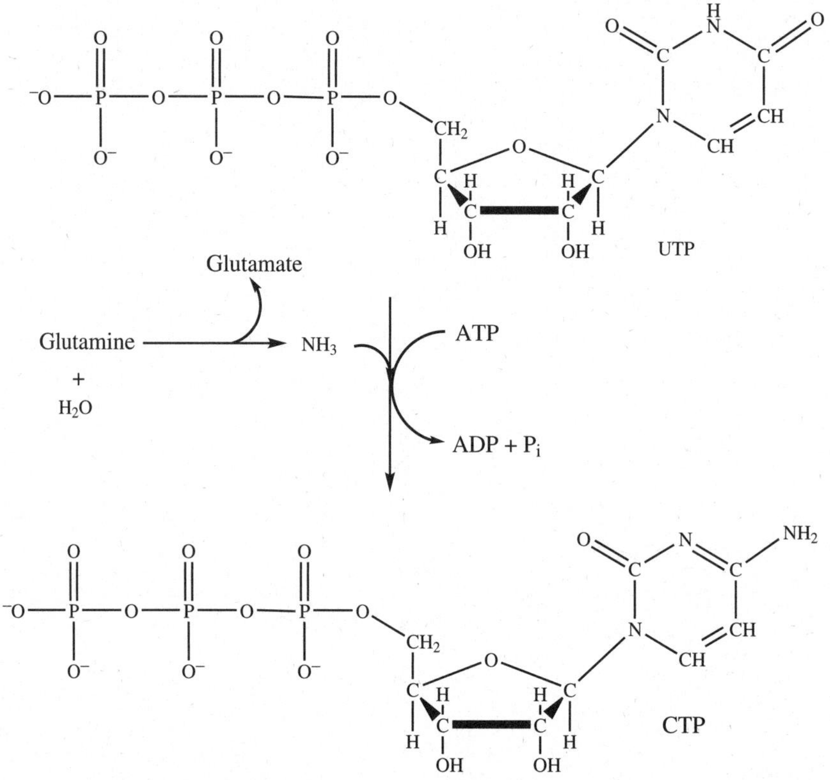

Figure 14-9:
Conversion
of UTP
to CTP.

Back to the Beginning: Catabolism

Catabolism, remember, is the breaking down of molecules to provide energy. In many cases, a complete breakdown isn't necessary because the products from a partial breakdown can be reused when necessary.

Nucleotide catabolism

The breakdown of the nucleotides begins with the removal of a phosphate group (from C-5). Next, a phosphate attaches to C-1 to give the sugar-1-phosphate, and the base is left. In humans and many other species, uric acid (see Figure 14-10) is the product of further degradation of purines. Other biochemical species further degrade uric acid into other products.

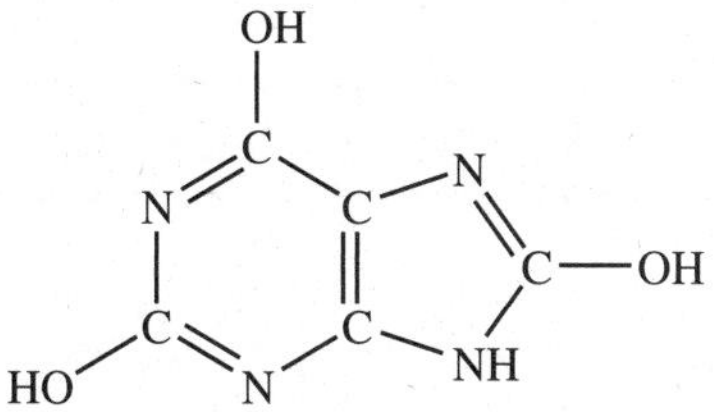

Figure 14-10:
Structure of
uric acid.

Amino acid catabolism

Hydrolysis of proteins yields the separate amino acids. Recycling these amino acids is possible (recycling is a good thing), as is using them in the synthesis of other amino acids or producing energy from them. Through *transamination* it's possible to transfer an amino group from any amino acid (other than lysine, proline, and threonine) and an α-keto acid. The general category of enzymes that catalyze this reaction is a *transaminase;* Figure 14-11 shows the general reaction. Nitrogen destined for elimination transfers to α-ketoglutarate to form glutamate. Transamination is important in the biosynthesis of alanine, aspartate, and glutamate.

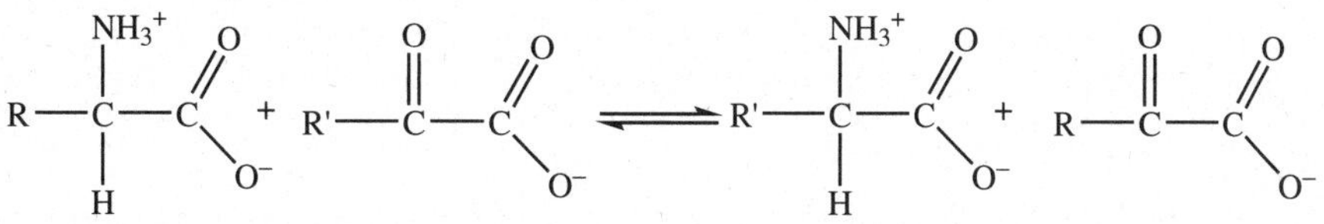

Figure 14-11:
General
trans-
amination
reaction.

Amino acid   α-keto acid   New amino acid   New α-keto acid

Oxidative deamination of glutamate forms α-ketoglutarate (to be recycled), an ammonium ion (to enter the urea cycle), and, indirectly, 3 ATP. Glutamate dehydrogenase and either NAD^+ or $NADP^+$ are necessary for this.

The deaminated amino acid (α-keto acid) may be further broken down to pyruvate or some other material the body can use to form glucose. These acids are called *glucogenic*. The alternative is to break down the α-keto acid to acetyl CoA and acetoacetic acid. These acids are called *ketogenic*. To further confuse you, some amino acids may be both glucogenic and ketogenic (see Table 14-2). These are the two possible fates of the carbon skeleton of the amino acids. The degradation of the amino acid transforms the carbon skeletons into intermediates in the citric acid cycle or into materials convertible to glucose.

Table 14-2	Glucogenic and Ketogenic Amino Acids
Glucogenic	Alanine, arginine, asparagine, aspartate, cysteine, glutamate, glutamine, glycine, histidine, methionine, proline, serine, threonine, valine
Ketogenic	Leucine
Both	Isoleucine, lysine, phenylalanine, tryptophan, tyrosine

The general process for amino acid catabolism is cyclic, with the various amino acids entering at different points. Figure 14-12 shows the basic scheme.

Heme catabolism

The other important nitrogen compound in red-blooded organisms is *heme*. This species occurs in both hemoglobin and myoglobin. Hemoglobin is released as aged red blood cells are destroyed. The globin portion hydrolyzes to the appropriate amino acids. The iron separates from the heme and is stored in *ferritin*. Through a series of steps, *bilirubin* forms from the heme. The gallbladder temporarily stores bilirubin until the organism eliminates it.

A similar process occurs during bruising. A trauma causes a rupturing of the blood vessels and blood pools. Large bruises may change color because of the breakdown of hemoglobin from the pooled blood cells. The noticeable colors of a bruise are caused by the sequential breakdown of hemoglobin (which produces a red-blue color) to biliverdin (green) to bilirubin (yellow) to hemosiderin (golden brown). As the body reabsorbs these products, the bruise fades.

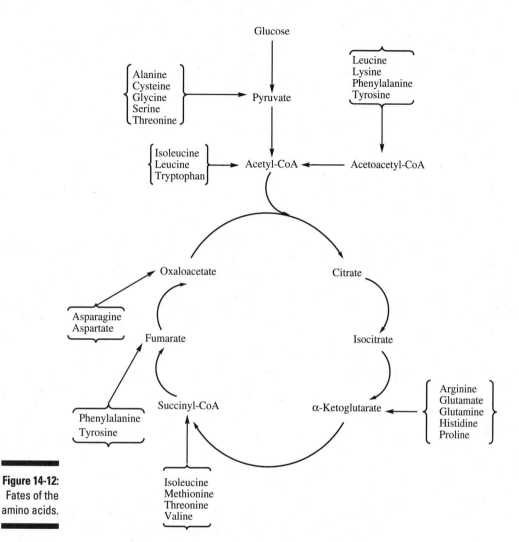

Figure 14-12:
Fates of the
amino acids.

Process of Elimination: The Urea Cycle

The catabolism of nitrogen-containing compounds yields recyclable nitrogen compounds and ammonia. Glutamine serves as temporary storage and transportation of the nitrogen. However, because even small amounts of ammonia are toxic to humans, ammonia must be converted to a less toxic form for elimination. The first step involves the conversion of ammonia, as the ammonium ion, to carbamoyl phosphate. The enzyme utilized for this conversion is *carbamoyl phosphate synthetase*. Figure 14-13 illustrates this reaction.

$$2\ ATP \qquad 2\ ADP$$

$$NH_4^+ + CO_2 \longrightarrow NH_2-\overset{\overset{O}{\|}}{C}-O-\overset{\overset{O}{\|}}{\underset{\underset{O^-}{|}}{P}}-O^- + P_i$$

Carbamoyl phosphate

Carbamoyl phosphate enters the urea cycle by joining to *ornithine* to produce *citrulline,* with the enzyme *ornithine transcarbamoylase* catalyzing this reaction. The enzyme *arginosuccinate synthetase,* with energy from the hydrolysis of ATP, joins aspartate to citrulline to form *arginosuccinate. Arginosuccinase* then catalyzes the splitting of arginosuccinate to *fumarate* and *arginine.* The enzyme *arginase* completes the cycle by cleaving arginine into urea (for elimination) and ornithine (for recycling). Figures 14-14 and 14-15 show the urea cycle and the compounds involved in it.

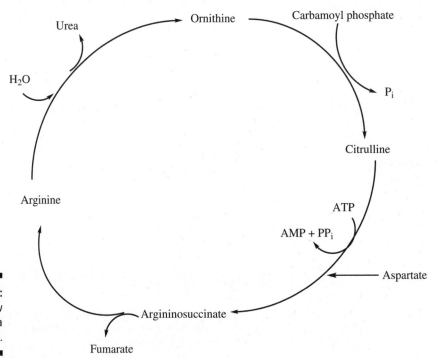

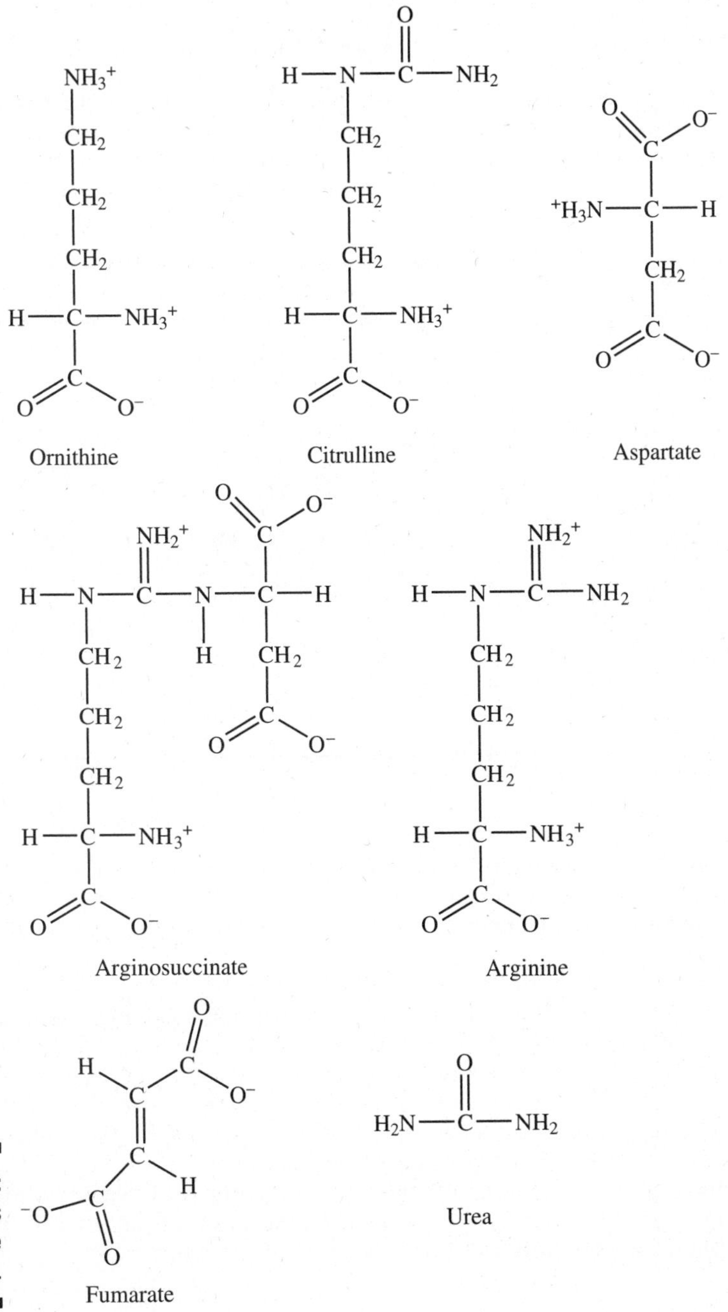

Figure 14-15: Compounds from the urea cycle.

Amino Acids Once Again

The synthesis of proteins requires 20 amino acids. Humans can synthesize 10 of these amino acids if they aren't readily available. These are the *nonessential* amino acids. The remaining 10 amino acids, the *essential* amino acids, must come from the diet. Table 14-3 summarizes these amino acids.

Table 14-3	Essential and Nonessential Amino Acids
Essential Amino Acids	*Nonessential Amino Acids*
Arginine*	Alanine
Histidine	Asparagine
Isoleucine	Aspartate
Leucine	Cysteine
Lysine	Glutamate
Methionine	Glutamine
Phenylalanine	Glycine
Threonine	Proline
Tryptophan	Serine
Valine	Tyrosine

* Not essential in adults

A *complete protein* supplies all essential amino acids. Not all proteins are complete — many are *incomplete proteins*. In order to avoid disorders caused by amino acid deficiencies, humans should eat a diet that contains complete proteins.

Transamination is important in the biosynthesis of alanine, aspartate, and glutamate. Converting aspartate to asparagine and glutamate to glutamine is easy. The synthesis of proline requires four steps, beginning with glutamate. The synthesis of serine begins with the glycolysis intermediate 3-phosphoglycerate, and after three steps, serine forms. Converting serine to glycine is easy. If sufficient phenylalanine is available, the catalyzed hydroxylation converts it to tyrosine. If sufficient methionine is available, the body can convert some of the excess to cysteine. Arginine comes from the urea cycle, but infants don't get sufficient quantities from this source.

Metabolic Disorders

When something is out of whack with an organism's metabolism, problems arise that must be treated. We discuss some of these disorders in this section.

Gout

Gout is the result of overproduction of uric acid, which leads to the precipitation of sodium urate in regions of the body where the temperature is lower than normal (98.6 degrees Fahrenheit, or 37 degrees Celsius). These low temperature regions are commonly found in the joints of the extremities. Sodium urate may also precipitate as kidney stones. Treatment is partly dietary and partly pharmaceutical. Dietary restrictions include limiting the intake of alcohol and foods high in nucleic acids (meats), which aggravate the conditions. Doctors often prescribe drugs that inhibit the enzyme that produces uric acid.

Gout may also be the result of faulty carbohydrate metabolism. A deficiency in glucose-6 phosphatase forces phosphorylated carbohydrates to form ribose 5-phosphate instead of glucose. Excess ribose 5-phosphate leads to excess PRPP, which, in turn, stimulates the synthesis of purines. The excess purines cause the production of more uric acid.

Lesch-Nyhan syndrome

Lesch-Nyhan syndrome is another example of defective purine catabolism leading to excess uric acid. Patients with this disorder normally excrete four to five times as much uric acid as gout patients do. This is a genetic disease marked by a recessive X-linked trait; the trait is carried by the mother and is passed on to her son. This disease has no treatment at the present time.

Albinism

Albinism, a recessive trait, is an inborn error of tyrosine metabolism. *Tyrosine* is the precursor of *melanin,* the pigment responsible for hair and skin color. In at least one form of albinism, the problem appears to be due to a deficiency of the enzyme *tyrosinase.* A variation of albinism involves a temperature-sensitive form of tyrosinase. The enzyme is only effective at lower than normal temperatures, as found in the extremities. This form of tyrosinase is responsible for the coloration of Siamese cats.

Alkaptonuria

Alkaptonuria is a benign condition that manifests itself as a darkening of the urine. The condition is the result of a problem in the catabolic breakdown of phenylalanine and tyrosine. A defective enzyme leads to an accumulation, and subsequent elimination, of one of the reaction intermediates.

Phenylketonuria

Phenylketonuria, or PKU, is the result of a deficiency in the enzyme phenylalanine 4-monooxygenase, which results in a problem in phenylalanine metabolism. The consequence is an accumulation of phenylalanine in the blood. High levels of phenylalanine enhance transamination to form abnormally high levels of phenylpyruvate. High levels of phenylpyruvate damage the brains of infants with the condition.

The high levels of phenylalanine lead to competitive inhibition of the enzymes responsible for melanin production from tyrosine. Because little tyrosine converts to melanin, afflicted infants have light blonde hair and fair skin (similar to albinism).

Early diagnosis in infants is important to prevent brain damage. Most states now require a blood test for PKU for all newborns. One test for PKU is to add $FeCl_3$ to the patient's urine. Phenylpyruvate reacts with iron ions to produce a green color. Another test is to assay for phenylalanine 4-monooxygenase activity. Treatment consists of maintaining a diet low in phenylalanine until at least the age of 3.

Part V
Genetics: Why We Are What We Are

By Rich Tennant

"The test results indicate that you're the descendant of a parade float."

In this part . . .

We roll up our sleeves and return to the subjects of genes and DNA to look at them much more closely. We cover the way DNA replicates itself and describe a number of applications related to DNA sequencing. Then it's off to RNA transcription and protein synthesis and translation.

Chapter 15

Photocopying DNA

In 1958, Francis Crick postulated what became the "central dogma of molecular biology." In this postulate, he, and later others, reasoned that DNA was the central source of genetic information and that it passed on some of this information to form RNA, which, in turn, passed on this information to form proteins. This central dogma is an extension of the one-gene one-protein hypothesis. DNA must be able to pass on its information both to later generations *(replication)* and to RNA *(transcription)*. RNA must finish the series by forming the appropriate proteins *(translation)*.

Some RNA, especially some viral RNA, can undergo replication and even reverse-transcription. Thus, RNA can produce both RNA and DNA. Genetic researchers initially thought this ability was in conflict with the central dogma, but Crick reasoned that RNA creating DNA was an extension of this postulate.

Many of the viruses capable of reverse-transcription can cause cancer.

In this chapter, we examine the replication process in detail, showing you the process by which DNA reproduces itself and passes along the genetic code.

Let's Do It Again: Replication

The primary structure of DNA consists of two polynucleotide strands held together by hydrogen bonds. Adenine forms hydrogen bonds to thymine, and cytosine forms hydrogen bonds to guanine (see Figure 15-1). The sequence of nitrogen bases contains the genetic information. The DNA molecules wrap around a protein called a *histone;* the combination of eight histones with the associated DNA is a *nucleosome.* (We talk more about histones in Chapter 16.)

A *gene* is a portion of a DNA molecule that carries specific information. The portion of the gene that codes for that specific information is called an *exon*. The portion of the gene that doesn't code for specific information is an *intron*.

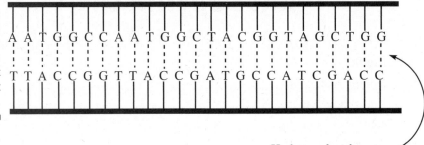

Figure 15-1:
A schematic illustration of the base pairs present in a segment of DNA.

Hydrogen bonds

Replication (illustrated in Figure 15-2) is the process that produces new copies of DNA molecules. One DNA molecule unwinds and opens, kind of like a zipper, and produces two exact copies of DNA molecules. (Zippers and replication — we're not going there.) New nucleotides bind to the backbone of each strand of the opened DNA by forming hydrogen bonds to the nucleotides (the zipper's "teeth") that are already present. The process proceeds along the opening DNA strand until each half of the original DNA has a complementary strand of hydrogen bonded to it. The result is two DNA double helixes, each with half old DNA and half new. This process doesn't sound like much fun, but it works for DNA.

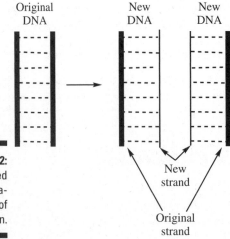

Original
DNA

New
DNA

New
DNA

New
strand

Original
strand

Figure 15-2:
A simplified representation of replication.

Because of the specific hydrogen bonding, the new strands contain a nucleo-tide sequence that's complementary to the old strand's nucleotide sequence. Therefore, an exact duplicate of the original DNA can be created.

This description of replication is a simplification. It barely scratches the sur-face of this complicated process, but it gives you enough background infor-mation to understand what comes next.

The first step in understanding replication was the discovery of DNA poly-merase from _Escherichia coli,_ better known as the beloved bacteria, _E. coli._ Subsequent studies showed that this enzyme needed a DNA template and all four deoxyriboside triphosphates (dATP, dCTP, dGTP, and dTTP). In addition, a short section of RNA called a _primer_ is also needed. The enzyme prefers a single DNA strand for the template in order to produce a comple-mentary strand.

During replication, simultaneous duplication of the two DNA strands occurs (see Figure 15-3). Because the two strands of DNA are antiparallel, the mode of synthesis is different for each strand, but the overall process is the same: moving from one end to the other. For one strand the synthesis is from 5' → 3'. On the other strand it appears to be from 3' → 5', but in actuality it's also 5' → 3'. The 3' → 5' strand has a complication that we discuss later in this section.

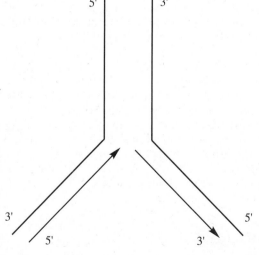

Figure 15-3:
A simplified scheme of the replica-tion of DNA.

The initiation of replication begins at a particular site, and after it's initi-ated, a series of fragments form discontinuously along one strand and con-tinuously along the other strand. The discontinuous fragments, known as _Okazaki fragments,_ contain from 1,000 to 2,000 nucleotides. The synthesis of the fragments is always in the 5' → 3' direction. Figure 15-4 illustrates in more detail how this overall process occurs.

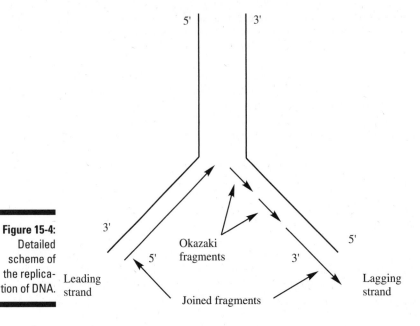

Figure 15-4:
Detailed
scheme of
the replica-
tion of DNA.

Researchers unexpectedly found that RNA synthesis is a prerequisite for the replication of DNA. Initially, an RNA primer, typically 20 to 30 nucleotides in length, forms on a single DNA strand. After it forms, deoxyribonucleotides add to the 3' terminus. Later, the RNA primer is removed and the appropriate DNA fragment is attached to produce the completed DNA.

At least a portion of the double-stranded DNA must be separated before replication can occur, and the separated portions can serve as templates. Enzymes known as *helicases* are responsible for this separation. Researchers don't understand the mechanism of separation very well and are still investigating it. Apparently, the helicase binds more strongly to one strand of the DNA than the other so that the enzyme squeezes in and pushes the other strand away. ATP hydrolysis provides the energy necessary to cause the enzyme to move along the one strand, nucleotide by nucleotide, a process that results in regions of the DNA opening like the zipper we talk about previously.

DNA polymerases

DNA polymerases are the enzymes responsible for joining the nucleotide triphosphate fragments to produce a strand of DNA, acting as the bricklayers and carpenters in its construction. This process only occurs in the presence of a DNA template (parent DNA). Before the enzyme can connect a nucleotide, the nucleotide must bind to the appropriate site on the template.

A cell may have more than one DNA polymerase. For example, in *E. coli,* three different enzymes perform the task of joining the nucleotide triphosphate fragments. These enzymes may also act as *exonucleases,* which have the opposite function of a polymerase; that is, they remove nucleotides from the DNA strand.

The addition of the nucleotides is always to the 3' end of a polynucleotide chain. DNA polymerases can't start building a nucleotide from scratch; a polynucleotide must already be present. In contrast, RNA polymerase *can* begin from scratch. RNA polymerase generates the RNA primer, using ribonucleotides, at the beginning of replication. DNA polymerase then takes over the task and adds deoxynucleotides to the RNA primer. The polymerization requires the presence of two metal ions to enable the joining of the nucleotide to the polynucleotide.

Replication of DNA needs to be error-free to ensure proper transmission of genetic information, and DNA polymerases are extremely effective in reducing errors. The enzyme binds tightly to the template and to the incoming nucleotide. This nucleotide is initially bound to the template through hydrogen bonding. If the wrong nucleotide is present, the subsequent binding to the polymerase is ineffective, and the nucleotide is "rejected." In addition to this checking, DNA polymerase also proofreads the preceding nucleotide to make sure it's correct. If the wrong nucleotide is present, it doesn't fit properly, and the erroneous nucleotide must be removed from the polynucleotide so that the correct nucleotide may enter. The exonuclease portion of the polymerases performs this function. The polymerase proofreads the polynucleotide chain as polymerization proceeds. Proofreading is in the reverse direction (3' → 5'). A nucleotide must already be in place before the polymerase can proofread. (We hope that our proofreader is as good as the DNA polymerases.)

The current model of DNA replication

In vitro studies show that in *E. coli,* replication begins when a protein binds a region of the DNA that contains four specific binding sites. This is the *origin of replication* site. After this protein binds, a helicase enzyme attacks the DNA and begins to unwind and separate the two strands. A third protein enters and holds the DNA strands open so that replication can continue. This third protein is the single-strand binding protein. The partially opened DNA and associated proteins are called the *prepriming complex* (see Figure 15-5).

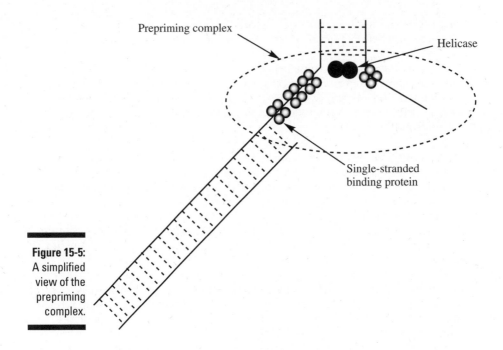

Prepriming complex

Helicase

Single-stranded
binding protein

Figure 15-5:
A simplified
view of the
prepriming
complex.

The DNA templates must be exposed in this manner. A DNA strand may have more than one origin of replication site; this allows replication to occur in many places at one time. Simultaneous replication allows the cell to replicate the entire strand in less time.

Replication can't continue until the exposed template is primed. A type of RNA polymerase known as *primase* binds to the prepriming complex in a region known as the *primosome*. Primase synthesizes a short RNA segment of about five nucleotides. Primase is capable of performing this function because its proofreading ability isn't as efficient as that of DNA polymerase. For this reason, a nucleotide doesn't need to already be present to be checked. Because the primer consists of ribonucleotides instead of deoxyribonucleotides, it is temporary and is detected and removed later. After it's removed, the appropriate deoxyribonucleotides join to complete the DNA strand (see Figure 15-6).

Although both strands of DNA serve as templates, the replication process differs on each strand. The point where the strands split and replication occurs is the *replication fork* or *replication bubble*. Because the two strands are antiparallel, and DNA polymerase only works in the 5' → 3' direction, direct replication only works on one strand, which is called the *leading strand*. The other strand is the *lagging strand*. (Sounds like me [John] walking my dogs — the Lab leads and the schnauzer lags.)

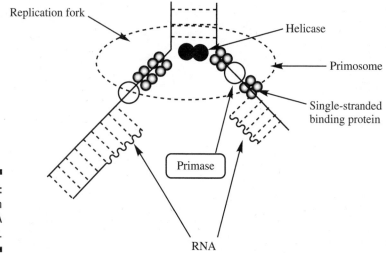

Helicase

Primosome

Single-stranded
binding protein

Primase

RNA

Figure 15-6:
Formation
of the RNA
primer.

As the DNA strands separate, eventually there's enough room for the molecular machinery to begin synthesis in the reverse direction on the lagging strand. (The reverse direction on the antiparallel lagging strand is still $5' \rightarrow 3'$.) Replication on the lagging strand is discontinuous, and fragments of about 1,000 nucleotides form, called, as we note earlier, Okazaki fragments. DNA ligase then joins the fragments to produce a continuous strand.

DNA polymerase III holoenzyme (complete enzyme) simultaneously produces DNA on both the leading and lagging strands, though the mechanisms on the two strands are different. On the leading strand the process is continuous, whereas on the lagging strand it's discontinuous and more complex. To carry out the polymerization on the lagging strand, this strand loops around so that polymerization in the $5' \rightarrow 3'$ direction can take place. After about 1,000 nucleotides — an Okazaki fragment — the polymerase releases the loop and begins a new loop and fragment. Each Okazaki fragment has an RNA primer. DNA polymerase I synthesizes DNA in the gaps between the fragments and removes the primer section. DNA ligase then joins the fragments (see Figure 15-7). Wow! John wishes the carpenters who built his new house were that efficient!

The ends of the DNA strands require a different procedure than does the majority of the strand, and this procedure is especially important on the lagging strand. If care isn't taken, each replication cycle would result in a shorter DNA strand, eventually leading to the loss of important genetic material. To resolve this problem, the ends of the DNA strands contain *telomeres* — DNA segments that contain hundreds of repeating units. In humans, the repeating units are the hexanucleotide AGGGTT. The enzyme *telomerase,* in humans, detects the primer sequence GGTT and repeatedly attaches the hexanucleotide units, completing the DNA strand.

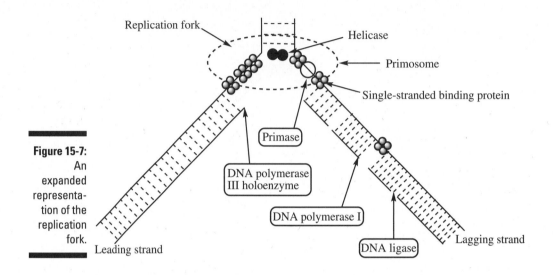

Figure 15-7:
An
expanded
representa-
tion of the
replication
fork.

Mechanisms of DNA repair

All cells have a variety of DNA repair mechanisms, which are necessary to repair defective DNA and ensure retention of genetic information. Damage to DNA may occur during replication or by the action of radiation or chemicals. A rare error known as *xeroderma pigmentosum* impairs these repair mechanisms. Individuals who suffer from it are extremely susceptible to cancers, especially skin cancers. Eventually, the skin cancers metastasize, leading to death.

Telomeres and aging

Each chromosome has a repetitive DNA sequence at the end known as a telomere. This sequence serves to protect the end of the chromosome from degradation. Part of the reason why telomeres are necessary is because as replication proceeds in the 5′ to 3′ direction, a point is reached where there is no longer room for a primer and it's not possible to replicate the last few nucleotides. This results in the loss of the end of the DNA strand. What is lost is part of a telomere, not important genetic information. An enzyme, telomerase, reverses transcriptase and thus reverses the telomere loss.

During an organism's lifetime, more telomeres are lost than are replaced. (In a few cells, like white blood cells, germ cells, and stem cells, telomerase activity is high.) The gradual shortening of the telomeres limits how many times a cell can divide. This suggests a mechanism for aging and places a limit on the life span of the organism. Organs containing a high proportion of cells that are no longer capable of division can't function properly or repair themselves.

Telomeres also inhibit the chromosomes from fusing or rearranging. Either of these abnormalities may lead to cancer. Rapid cell division in cancer cells can lead to early death of the cancer cells. In some cases, cancer cells are capable of maintaining the length of their telomeres, making the cancer cells "immortal."

The three general types of repair mechanisms are

- ✔ Direct repair
- ✔ Base-excision repair
- ✔ Nucleotide-excision repair

One example of damage needing repair is the formation of a thymine dimer (see Figure 15-8) by ultraviolet (UV) light. The *thymine dimer* is an example of a pyrimidine dimer, and its presence causes distortion of the DNA in the region. Thymine (a pyrimidine) should base-pair with adenine (a purine). Other problems include base mismatches and missing or additional bases.

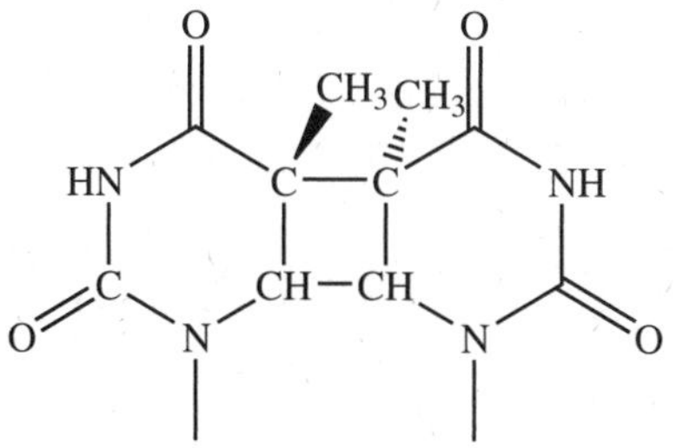

Figure 15-8: Structure of a thymine dimer.

Direct repair

In direct repair, the correction of the problem occurs in place. The photo-reactivating enzyme, DNA photolyase, binds to the cyclobutane ring present in a thymine dimer, using light energy to cleave this dimer into the original bases.

Base-excision repair

In base-excision repair, the correction of the problem involves removal and replacement of the base. This is necessary whenever a modified base is present. Modified bases have various causes, such as radiation or certain chemicals. The presence of a modified base normally results in a recognizable distortion in the DNA molecule. An enzyme, behaving as a glycosylase, cleaves the glycosidic bond to release the base from the deoxyribose. The result is an AP site, *AP* meaning *apurinic* or *apyrimidinic.* In an apurinic site, the purine base is absent; in an apyrimidinic site, the pyrimidine base is absent. An AP endonuclease recognizes this site and cuts the DNA backbone adjacent to the site. Next, a deoxyribose phosphodiesterase completes the removal of the remaining deoxyribose phosphate. DNA polymerase I then inserts a replacement nucleotide to base pair with the nucleotide in the complementary DNA strand. Finally, DNA ligase connects the units to yield the repaired strand — kind of like an electrician cutting out a bad circuit and splicing a good one in its place.

Nucleotide-excision repair

In nucleotide-excision repair, the correction of the problem involves the removal of a segment of DNA around the problem followed by its replacement. When this mechanism occurs, DNA sequences on both sides of the error are cut from the DNA strand. Typically, an exonuclease removes a 12-nucleotide section. DNA polymerase I then synthesizes a replacement segment of the strand, and DNA ligase then finishes the repair.

Mutation: The good, the bad, and the ugly

Several types of mutations are known. DNA repair mechanisms try to prevent new mutations, but such mechanisms aren't always effective. Known mutations include the substitution of one base pair for another, the insertion of one or more base pairs, and the deletion of one or more base pairs. Changes, especially subtle ones, may occur during or after replication.

The substitution of one base pair for another is a common mutation. The substitution takes two forms: transition and transversion. In a *transition,* a purine replaces the other purine (see Figure 15-9) or a pyrimidine replaces the other pyrimidine (see Figure 15-10). In a *transversion,* a purine replaces a pyrimidine or vice versa.

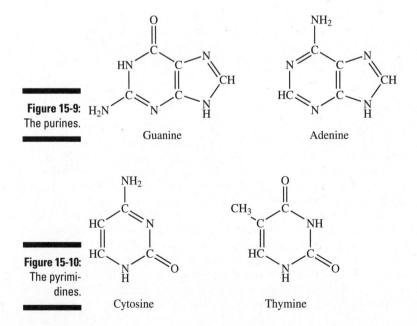

Figure 15-9: The purines.

Guanine   Adenine

Figure 15-10: The pyrimidines.

Cytosine   Thymine

Any uncorrected discrepancy in the genetic code becomes recognized as "normal" in all future generations. This new genetic code is a *mutation.* The change in the base sequence may or may not affect the amino acid for which the codon codes. For example, changing from GTT (coding for leucine) to GTG (also coding for leucine) results in no change. However, if the change results in coding for a different amino acid, the resultant protein functions differently. If the new protein exhibits improved function, the organism benefits from the change. But if the new protein exhibits impaired function — the more likely situation — the organism suffers from the change. Problems from impaired function are genetic diseases. Table 15-1 lists some of these.

Table 15-1	Some Genetic Diseases in Humans
Disease	*Defective Protein*
Acatalasia	Catalase
Albinism	Tyrosinase
Cystic fibrosis	CF transmembrane conductance regulator
Fabry's disease	α-Galactosidase
Gaucher's disease	Glucocerebrosidase
Goiter	Iodotyrosine dehalogenase
Hemochromatosis	Hemochromatosis
Hemophilia	Antihemophilic factor (factor VIII)
Hyperammonemia	Ornothine transcarbamylase
McArdle's syndrome	Muscle phosphorylase
Niemann-Pick disease	Sphingomyelinase
Phenylketonuria	Phenylalanine hydroxylase
Pulmonary emphysema	α-Globulin of blood
Sickle cell anemia	Hemoglobin
Tay-Sachs disease	Hexosaminidase A
Wilson's disease	Ceruloplasmin (blood protein)

Restriction enzymes

Although not directly related to replication, restriction enzymes are important tools in genetic research. *Restriction enzymes,* or *restriction nucleases,* are capable of cutting DNA into fragments. Restriction enzymes were first found in prokaryote cells like *E. coli,* in which these enzymes locate and destroy invading DNA, such as that of a bacteriophage, but leave the cell's own DNA alone. Recent research focuses on the fact that these fragments can be manipulated so that DNA ligases can join the fragments into new DNA. Restriction enzymes are important in vitro biochemical tools that act as very accurate molecular scalpels. Cleavage may leave both DNA strands of equal length or one strand may be longer than the other (a *staggered* cut).

Researchers have identified more than 100 restriction enzymes. These enzymes recognize specific regions in the DNA and cleave DNA molecules into specific fragments. Because these fragments are smaller than the parent DNA is, they're easier to manipulate and analyze. Testing a strand of DNA with a series of restriction enzymes can provide a fingerprint of cleaved fragments. In fact, you can map the structure of DNA.

Many times in reading descriptions of genetic determination and modification, you run across the terms *in vivo* and *in vitro*. *In vivo* means in the cell, whereas *in vitro* means in a test tube.

Mendel Rolling Over: Recombinant DNA

Recombinant DNA technology allows the synthesis of DNA strands that contain one or more genes not originally present. The addition of new genes enables an organism to produce new biochemicals. For example, *E. coli* has been engineered to produce human insulin. Recombinant DNA technology also allows biochemists to add a gene to compensate for a defective gene.

Restriction enzymes are capable of removing DNA fragments of interest. For replication to occur, one of these fragments must join (or recombine with) another DNA strand. The DNA to which the fragment of interest is attached is the *vector.* Common vectors include *plasmids,* which are a naturally occurring DNA circle. The first step in adding the fragment is to create a staggered cut in the DNA of the vector. The longer end of the staggered cut is a "sticky" or cohesive end to which any DNA fragment can attach if it has the complementary DNA sequence. The complementary sticky end is present if the same restriction enzyme is used to excise the fragment of interest. DNA ligase completes the joining of the fragment to the vector (see Figure 15-11).

A DNA linker can bond to a DNA molecule to make it susceptible to a particular restriction enzyme. By this method, the cohesive ends characteristic of any restriction enzyme may be added to almost any DNA molecule. The completed DNA can undergo replication.

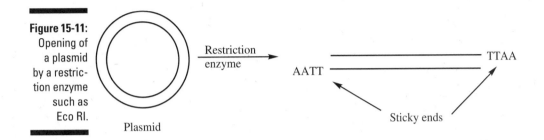

Figure 15-11:
Opening of a plasmid by a restriction enzyme such as Eco RI.

Plasmids are, to a certain extent, accessory chromosomes. They can replicate independently of the host chromosomes. Thus, a cell may have multiple copies of a particular plasmid. This replication, in general, makes plasmids more useful as vectors than host chromosomes. Thus far, these plasmids have been shown to be relevant only in bacterial organisms.

The addition of "new" genes to an organism produces an organism that's *transgenic* and could even be considered a new species. These organisms could infect humans and lead to a new disease that has no known treatment. To minimize potential risks posed by these organisms, researchers use either enfeebled (weakened) organisms or ones that don't infect humans.

Patterns: Determining DNA Sequences

Restriction enzymes are a major tool in the determination of the base sequence in DNA. The cleaved DNA fragments are significantly smaller than the parent DNA, making manipulation and analysis significantly easier. To separate the fragments after cleavage, gel electrophoresis is often used, which then allows for further analysis.

Getting charged up about gel electrophoresis

Gel electrophoresis (see Figure 15-12) is a biochemical technique used to separate and purify proteins and nucleic acids that differ in charge, size, or conformation. The sample is placed into wells within a gel — a polymer that's specifically formulated for the type of analysis. These gels are in the shape of a thin slab and have the consistency of Jell-O, though they probably don't taste nearly as good. The gels act as a molecular sieve. To separate proteins or small nucleic acids (DNA, RNA, and so on), biochemists use cross-linked polyacrylamide. To separate larger nucleic acids, biochemists use agarose, an extract from seaweed.

The gel is immersed in a buffer solution, and an electrical current is applied to the ends of the gel. The charged species within the sample migrate toward one or the other of the electrodes. Proteins may have either a positive or a negative charge, but at the proper pH, nucleic acids have only a negative charge. The positively charged species move toward the negatively charged end of the gel, and the negatively charged species move toward the positively charged end. Normally, a buffer adjusts the pH so that all the species of interest have either a positive or negative charge.

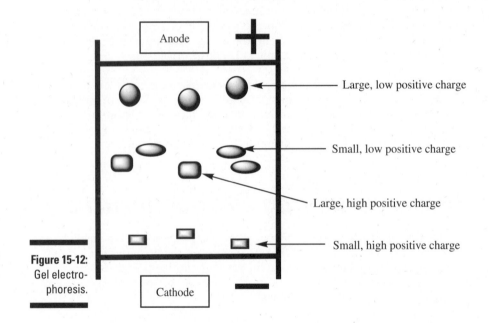

Figure 15-12:
Gel electro-
phoresis.

Different molecules move at different speeds through the gel. When the smaller, faster molecules have about reached the end of it, the process is stopped, and the molecules are stained to make them visible. Sometimes, scientists add agents to cause the molecules to fluoresce (glow) under UV light, and then they may take a photograph of the gel as it's exposed to the UV light. When several samples, including molecular weight markers of known sizes, are run side by side, scientists can determine the molecular weight of a sample component. This is one step in the identification of unknown components.

The separation of DNA fragments by gel electrophoresis readily distinguishes even minor differences between the fragments. Some gel concentrations are more useful in separating large fragments than small fragments. And some gels can help distinguish between fragments that only differ by one base in several hundred. Modification of the electrophoresis method provides further separation. Each type of DNA gives a different pattern, so distinguishing between two different samples is possible. Two samples that give identical patterns must be from the same source or from identical twins.

In the analysis and manipulation of genetic material, being able to identify whether a certain sequence of nucleotides is present is advantageous. The general method for finding a particular sequence of nucleotides in DNA was developed by Edwin Southern and is called *Southern blotting.* This method uses radioactive ^{32}P as a label that's easily detectable. This radioisotope is incorporated into the phosphate in some of the nucleotides. Determination of a particular nucleotide sequence in RNA is achieved through *Northern blotting,* and protein identification is achieved through *Western blotting.* (Unlike Southern, the names Northern and Western don't refer to persons; they're just analogous to Southern.) Alternatively, Southern, Northern, and Western blotting are DNA, RNA, and protein blots, respectively.

Determining the base sequence

Since the first isolation of DNA, a number of methods have been developed to determine the base sequence. In general, the *Sanger dideoxy method* has replaced all others. It employs the controlled termination of replication with modified nucleotides containing dideoxyribose in place of deoxyribose.

DNA fragments produced by employing restriction enzymes are denatured to give single-stranded DNA. (*Denaturing* typically involves heating a DNA-containing solution to 205 degrees Fahrenheit [96 degrees Celsius] for a few seconds.) Four samples of this DNA are treated separately to produce double-stranded DNA through replication, and each sample contains a small quantity of a different dideoxy nucleotide. The dideoxy nucleotide contains dideoxyribose. The absence of an additional oxygen atom in dideoxyribose means that no 3' hydroxyl group is available to continue replication. Thus, the incorporation of a dideoxynucleotide terminates the DNA chain. (See Figure 15-13.)

One of the four samples will contain a small quantity of the dideoxy analog of the nucleotide dGTP. This "defective" unit enters the new DNA strand as the complement to a cytosine base in the original fragment. Separation of the new material from the original strand material gives a set of DNA fragments of varying length. These fragments are then separated by electrophoresis according to length (size). The length of each of these fragments locates the position of each C in the original strand. The other three samples give the positions of all A, T, and G bases in the original strand.

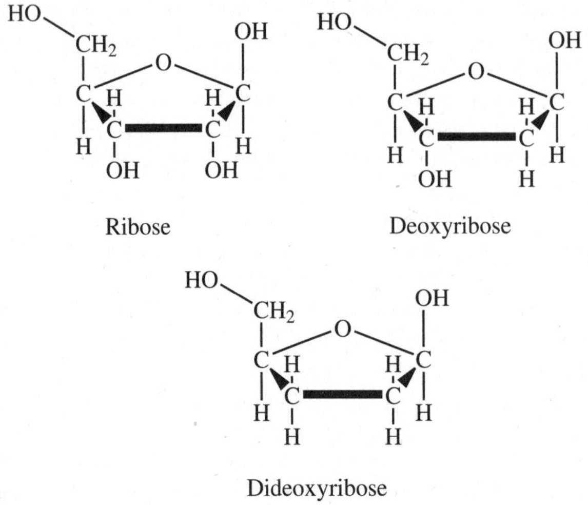

Ribose

Deoxyribose

Figure 15-13:
Structures
of ribose,
deoxyri-
bose, and
dideoxy-
ribose.

Dideoxyribose

Fluorescence tagging is a useful modification to this method. Each of the dide-oxy nucleotides has a different fluorescent tag attached. After attaching the tags, you can conduct all four experiments in one container. Separating the fragments by electrophoresis and examining the tags gives a colored pattern that shows all the bases in sequence. This method works for fragments of up to 500 bases.

To conduct these studies, you need a sufficient amount of genetic material. Lack of sufficient quantities of a sample has been a problem, especially with forensic evidence. Therefore, scientists developed ways of quickly duplicat-ing sufficient quantities of identical DNA fragments or producing a number of DNA strands from a very small sample. *Polymerase chain reaction* (PCR) is a useful method to amplify specific DNA sequences. PCR is an in vitro proce-dure that requires knowledge of the base sequences adjacent to a particular target sequence (namely the flanking sequences). However, you don't need to know the base sequence in the target region. Denaturation of a DNA sample provides two separate strands. Two primers are added to the mixture and one primer attaches to the flanking sequence of each strand. DNA polymerase begins replication starting at each of these primers. Repeating these steps quickly generates a large quantity of DNA. After 30 or so cycles, a billion-fold amplification occurs. Thirty cycles take less than one hour.

The butler did it: Forensic applications

Scientists can identify a species by the isolation and examination of the DNA sequences unique to that species. For example, DNA analysis is useful in the identification of organisms, such as bacteria, that may be polluting water,

food, and other samples. DNA analysis has been used to establish pedigrees
for livestock breeds as well as to identify endangered species in the prosecu-
tion of poachers. However, the application that has received the most public-
ity is in the area of *forensics*.

Because an individual's DNA comes from both the mother and father, the
DNA is unique to that individual (except in the case of identical twins). Even
brothers and sisters, including fraternal twins, show some variation in their
DNA. This fact makes DNA analysis very valuable in forensics investigations
(as anyone who's ever watched an episode of *CSI* or *Bones* can attest).

In order to identify an individual, forensic investigators examine 13 regions
(markers) of the DNA sample that vary significantly from individual to individual. Two individuals *could* have the same DNA pattern in these 13 regions,
but the likelihood is only about one chance in a billion. The investigation of
additional markers can improve the discriminating ability of the procedure.
Investigators then combine the results into a DNA profile — also known as a
DNA fingerprint — of the individual.

You can isolate DNA samples from blood, hair, bone, fingernails, teeth, and
any type of bodily fluid. In a typical crime scene analysis, samples are taken
from the evidence and suspects, and the DNA is extracted and analyzed for
the specific markers. A match of a single marker doesn't prove that an individual was at the crime scene, but the matching of four or five markers indicates
a very high probability that the person was present. PCR may be necessary if
the sample is very small (see the preceding section).

Methods of analysis

Several techniques are used in DNA analysis. The three most common are
RFLP, PCR, and STR. In RFLP *(restriction fragment length polymorphism),* the
DNA sample is digested with a specific enzyme, a restriction endonuclease. This enzyme cuts DNA at a specific sequence pattern. The presence or
absence of these sites in a DNA sample leads to variable lengths of DNA fragments. Gel electrophoresis then separates these fragments.

RFLP is one of the original forensic DNA analysis techniques. However, it
requires relatively large amounts of DNA, and samples contaminated with
dirt and mold are difficult to analyze with RFLP. It has been somewhat
replaced with PCR enhancement, followed by STR analysis.

Polymerase chain reaction or PCR (which we discuss in the earlier section
"Determining the base sequence") is a useful technique that reduces the
sample size requirement of RFLP; in essence it's a DNA amplifier. PCR quickly
makes millions of exact copies of the DNA sample. Using PCR, scientists can
do DNA analysis on a sample as small as a few cells and on samples that are
extensively degraded. After PCR treatment, they can analyze the sample with
RFLP or STR.

PCR has been a very useful tool for forensic scientists. Before PCR, the DNA sample to be analyzed had to be large enough to undergo testing. Using PCR, the sample size can be minute. The DNA can be amplified and then subjected to identification. The use of PCR has, in fact, freed some innocent people on death row; without PCR, they would have been executed.

In STR *(short tandem repeat)* analysis, the DNA sample is quickly examined for 13 specific regions. The FBI uses this standard STR profile in its CODIS (Combined DNA Index System) program, which links national, state, and local databases of DNA profiles from felons, missing persons, and unsolved crime scenes. CODIS has an index of more than 3 million DNA profiles.

Paternity testing

Along with crime scene analysis, paternity testing is one of the most widely used applications of DNA testing. The procedure begins with the collection of DNA samples from the mother, child, and alleged father(s). The DNA profiles of the child and mother are first determined. The markers not inherited from the mother must have come from the biological father. The alleged father's DNA profile is then compared to the child's. If the man's DNA profile contains markers common to the child but not to the mother, then the probability that he's the biological father is great. Figure 15-14 indicates that alleged father 2 is more likely to be the biological father than alleged father 1.

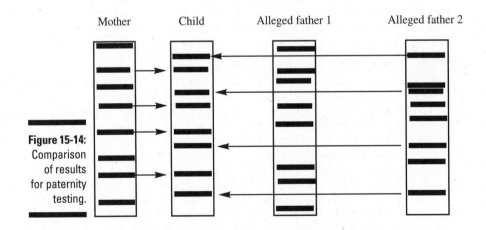

Mother   Child   Alleged father 1   Alleged father 2

Figure 15-14: Comparison of results for paternity testing.

Genetic Diseases and Other DNA Testing Applications

DNA testing always seems to find new ways of being useful. It has been used for a number of years, for example, in determining the gender of athletes. In addition to gender testing, the NFL used a strand of synthetic DNA to mark all the Super Bowl XXXIV footballs as a way to combat fraud associated with sports memorabilia. And a section of DNA was added to the ink used to imprint all official goods marketed at the 2000 summer Olympic Games. This same technology is used to tag original artwork and to track possibly poached ivory.

Genetic diseases are the result of an abnormal pattern in the DNA of an individual. These diseases are inherited, though some individuals are only carriers and not sufferers. Recently, quite a bit of research has been done to determine the genetic pattern that's causing diseases and to find ways to detect the probability of passing on a disease to offspring. However, methods of treatment for most all these diseases are limited. The dream of researchers is to find the means of correcting these genetic diseases through genetic modifications, now known broadly as *gene therapy*. Researchers have investigated several of these genetic diseases in detail. In this section we briefly examine a few of the more well-known genetic diseases.

Sickle cell anemia

Sickle cell anemia is an inherited genetic disease of the blood's *hemoglobin,* a component of red blood cells. Sickle cell anemia is the result of the change of a single amino acid in the protein sequence of hemoglobin. This change involves the substitution of valine (nonpolar) for glutamic acid (polar). The condition affects millions of people throughout the world, especially those whose ancestors came from Africa, South America, Cuba, Saudi Arabia, and a few other countries. In the United States, it affects about 72,000 people. Sickle cell occurs in about 1 in 500 African-American births and in about 1 in 1,200 Hispanic-American births.

Hemoglobin is responsible for carrying oxygen from the lungs to the cells. In an individual with sickle cell anemia, the defective hemoglobin molecules clump together, causing the red blood cells to assume a sickle shape, hence the name. These abnormal cells have trouble squeezing through small blood vessels, causing oxygen depletion in organs and extremities, along with episodes of pain. These sickle cells also have a much shorter lifetime in the body, leaving the individual with chronic anemia. Many states now test newborns for sickle cell disease.

Hemochromatosis

Hemochromatosis, one of the most common genetic diseases in the United States, is an inherited disease that causes the body to absorb and store far too much iron. This excess iron is stored in organs, such as the liver, pancreas, and skin (yes, the skin is considered an organ!). Hemochromatosis is caused by a mutation in the HFE gene, the gene that regulates the absorption of iron from food. If this defective gene is inherited from both parents, the person develops hemochromatosis. If the individual inherits the mutated gene from only one parent, the person is a carrier but won't necessarily develop the disease. About 5 Caucasian people in 1,000 carry both mutated genes, and 1 in 10 is a carrier. Genetic testing can detect hemochromatosis about 90 percent of the time.

Cystic fibrosis

Cystic fibrosis is a chronic and normally fatal genetic disease that affects the body's mucus glands. It targets the digestive and respiratory systems. About 55,000 people worldwide have cystic fibrosis. Most of them are Caucasians who have ancestors who came from northern Europe. One must inherit the mutated gene responsible for cystic fibrosis from both parents for the disease to appear. Estimates are that 1 in 20 Americans carry the abnormal gene. Most of these individuals aren't aware that they're carriers. Genetic testing is only about 80 percent accurate.

Hemophilia

Hemophilia is a genetic disorder caused by the lack of the blood-clotting factor stemming from a defective gene on the X chromosome. Females have two X chromosomes, so if one has a defective gene, chances are small that the other one is also defective. But such a female is a carrier. Males, however, have only one X chromosome, so if it's defective, the individual develops hemophilia. If a woman is a carrier, she has a 50 percent chance that her sons will have hemophilia and a 50 percent chance that her daughters will be carriers. Daughters of a hemophilic male are carriers. Genetic testing can detect the presence of the abnormal gene.

Tay-Sachs

Tay-Sachs is an inherited disease in which a fatty-acid derivative, a lipid called *ganglioside,* accumulates in the brain as a result of a mutation of a specific gene. Although this abnormal gene is found primarily in the Jewish population, some French Canadians and Louisiana Cajuns also carry it. The symptoms most commonly appear in infants. Death normally occurs before the age of 5. Although Tay-Sachs is a very rare disease, it's one of the first genetic diseases for which extensive and inexpensive genetic screening was developed. Screening tests were developed in the 1970s, and Israel offered free genetic screening and counseling. Because of this aggressive testing and counseling, the disease has been almost totally eradicated from Jewish families worldwide.

Ethics of genetic modification and testing

The emerging field of *bioengineering* has raised many ethical questions, including debates over stem-cell research, genetically modified crops, gender selection of children, genetic modification to enhance certain traits such as athletic ability, and so on. Public policy decisions related to cost are also being debated, as genetic modification and screening are generally expensive. Should these procedures be available only to those who can afford them, or should there be equal access? Are genetically modified crops safe? How should they be regulated? Many gray areas concerning genetic modification exist in the field of patent law. People have many questions and concerns but no quick answers.

Although the success in eradicating Tay-Sachs is directly related to genetic testing, such testing is not without its ethical questions. The major concern is one of privacy. DNA samples and profiles can be used to determine parentage and susceptibility to certain genetic diseases. Many people fear that the government, insurance companies, employers, banks, schools, and other organizations could use such information for genetic discrimination. In fact, in the United Kingdom, a man was denied treatment for hemochromatosis because his insurance company claimed it was a preexisting condition. Individuals applying for life insurance have reported other cases of genetic discrimination. Who gets to request the genetic screening, and who has access to the results? These are just a few of the questions we will be debating for many years to come.

Chapter 16

Transcribe This! RNA Transcription

In This Chapter

- Defining the role of RNA polymerase
- Examining transcription in prokaryotic and eukaryotic cells
- Breaking down the genetic code
- Regulating genes

In this chapter we concentrate on the synthesis of RNA, which is called *transcription* — the process whereby DNA produces RNA. First, a portion (a gene) of a DNA double helix opens. Nucleotides can then bind to the exposed DNA nucleotides through a process similar to replication. However, this process differs from replication in that only a portion of the DNA opens, and the entering nucleotides contain uracil in place of thymine. One gene yields one RNA molecule, which, in turn, may lead to the synthesis of one or more proteins.

The enzyme RNA polymerase joins the nucleotides to produce RNA in a process that occurs within the cell nucleus. The process begins as an initiation signal toward the 5' end of RNA and goes toward a termination sequence nearer the 3' end.

Types of RNA

Cells use several types of *ribonucleic acid,* or *RNA:*

- *Messenger RNA* (mRNA), biochemically a structure that isn't very stable, carries information from the cell nucleus into the cell's cytoplasm and must migrate to the ribosomes. Messenger RNA carries the actual genetic information necessary for the synthesis of a specific protein; however, the other forms of RNA are necessary to complete the process.

✔ *Transfer RNA* (tRNA) transfers amino acids to the ribosomes for protein synthesis. Transfer RNA is a relatively small form of ribonucleic acid, typically containing from 73 to 93 nucleotides.

✔ The relatively large *ribosomal RNA* (rRNA) resides in the ribosomes and has a direct influence on the synthesis of proteins. This form of RNA forms a complex with protein components and comes in three types (23S, 16S, and 5S); all three must be present in each ribosome.

✔ Finally, *small nuclear RNA* (snRNA) serves a number of ancillary functions.

RNA Polymerase Requirements

Three requirements are needed for the molecular machine, RNA polymerase, to operate. First, it requires activated precursors of each of the four *ribonucleoside triphosphates* (ATP, CTP, GTP, and UTP) from which to produce the new RNA (see Figure 16-1). Second, a divalent metal ion, either magnesium or manganese, is necessary. Finally, a template must be present. Single-stranded DNA works, but the preferred template is double-stranded DNA. However, the DNA strands must open (separate) to allow the RNA polymerase access.

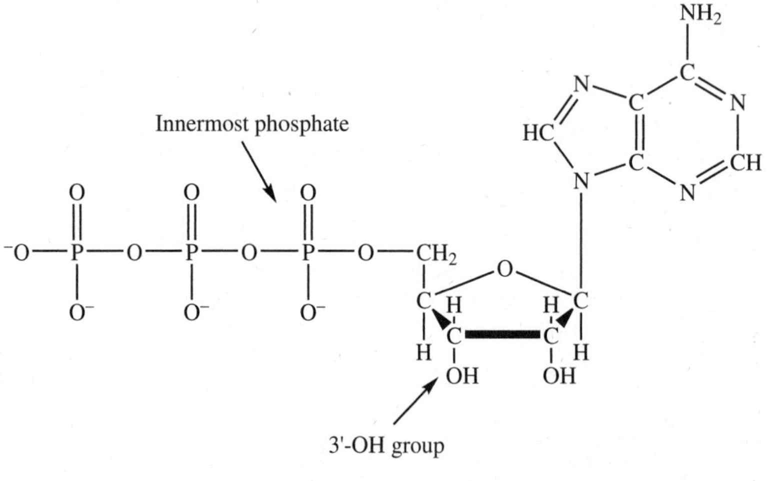

Figure 16-1:
Structure
of ATP.

Replication and transcription have many similarities. In both processes, the direction of synthesis is 5' → 3'. Elongation occurs as the chain's 3'-OH group attacks the innermost phosphate of the entering nucleoside triphosphate, a process called a *nucleophilic attack*. The hydrolysis of pyrophosphate provides the impetus to drive the process forward.

However, replication and transcription also differ. Unlike its DNA counterpart, RNA polymerase isn't capable of "reviewing" its work and then eliminating a mismatched nucleotide. RNA polymerase doesn't require a primer.

In simple organisms, such as *E. coli,* one type of RNA polymerase synthesizes all forms of RNA. More advanced organisms like human beings have different types of RNA polymerase. Usually, at least three different types are present in mammalian cells.

Making RNA: The Basics

The region of a DNA molecule that codes for a protein is a *structural gene.* Other regions are present to regulate this gene's activity. (We examine these regulatory regions later in this chapter.) To begin transcription, RNA polymerase must detect one particular gene present in a long DNA strand. Detection begins with the enzyme locating a region on the DNA strand known as a *promoter site,* which is upstream from the actual gene. (*Upstream* means on the 5' side.) RNA polymerase tightly binds to the promoter site, and after it's in place, transcription can begin.

Promoting transcription of RNA

In prokaryotic cells, the promoter sites are centered at –10 (the *Pribnow box*) and in the –35 region. The Pribnow box has the consensus sequence TATAAT centered at –10. The other site has the consensus sequence TTGACA. (Not all organisms have the same consensus sequence.) In eukaryotic cells, a promoter is centered at about –25 (the TATA box or *Hogness box*) or sometimes centered near –75 (the CAAT box). The consensus sequence in the Hogness box is TATAAA. The CAAT box has the sequence GGXCAATCT. In addition, eukaryotic genes may have enhancer sequences up to several thousand bases away from the start site and on either side (see Figure 16-2).

The position of sequences along the DNA chain starts at the beginning of a gene. The gene's first nucleotide is +1. Counting upstream (toward the 5' terminus) is negative. Thus, ten nucleotides before the beginning of the gene would be –10.

Transcription proceeds as an RNA polymerase moves along the DNA strand. Eventually, the enzyme encounters a termination signal. Prokaryotic cells have two termination signals. The first is a *base-paired hairpin,* which consists of a self-complementary sequence rich in C and G followed by a sequence of several instances of U. After the sequence forms, the new RNA detaches from the template. The other method uses a *rho protein.*

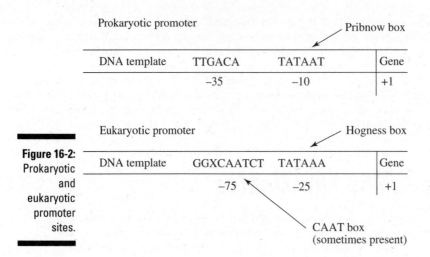

Figure 16-2:
Prokaryotic and eukaryotic promoter sites.

Biochemists don't understand the termination in eukaryotic cells very well. In eukaryotic cells, mRNA undergoes further modification after transcription. A "cap" is attached to the RNA's 5' end, and a poly(A) tail goes onto the other end. These modifications increase the lifetime of mRNA.

The stages in RNA synthesis are *initiation, elongation,* and *termination.* To accomplish these tasks, RNA polymerase must perform a series of functions. The enzyme must travel along a DNA strand until it encounters a promoter site. As it "sticks" to the promoter site, it unwinds a short segment of the DNA double helix and separates the strands to reach the template. Then the appropriate ribonucleoside triphosphate enters, and hydrolysis of the phosphate occurs in order to supply the needed energy.

Each ribonucleoside triphosphate is brought in as the RNA polymerase moves along the DNA strand. (The DNA unwinds as the enzyme passes and rewinds after the enzyme has passed.) This process continues until the RNA polymerase finds a termination signal. The enzyme also must interact with *transcription factors* or *trans-acting factors* — proteins that act as activators or repressors — to regulate the rate of transcription initiation.

The best understood operation of RNA polymerase comes from studies of the prokaryotic cells of *E. coli.* Eukaryotic cells behave in a similar, though more complicated, manner. One major difference between the two is that in pro-karyotic cells, transcription and translation (protein synthesis) may occur almost simultaneously, whereas in eukaryotic cells there's a gap between the two processes while the mRNA moves from the nucleus to the ribosome. The other major difference is that RNA in eukaryotic cells almost always requires processing after synthesis, whereas prokaryotic RNA is usually ready immediately after synthesis. Processing includes adding a cap, adding a poly(A) tail, and — in nearly all cases — splicing to remove *introns* (segments that don't code for protein).

Prokaryotic cells

RNA polymerase in *E. coli* contains four subunits that combine to form a holoenzyme designated $\alpha_2\beta\beta'\sigma$. The purpose of the σ subunit is to help find the promoter and to help initiate RNA synthesis. After synthesis begins, this unit leaves the remainder, the core enzyme. The catalytic site in the core enzyme contains two divalent metal ions, one that stays with the core and one that enters with the ribonucleoside triphosphate and leaves with the cleaved pyrophosphate. Three aspartate residues aid in the binding of the metal ions. Although DNA polymerase and RNA polymerase have very different overall structures, their active sites are similar.

In the absence of the σ subunit, RNA polymerase would bind tightly to DNA at any point. When the σ subunit is present, binding at other than a promoter site is significantly lower. Due to its reduced affinity, the holoenzyme can slide along the DNA strand until a σ subunit detects a promoter site. RNA polymerase binds to this site more strongly than to other positions on the DNA strand. The efficiency of this binding is one form of regulation. A number of σ subunits are present; each is designed to recognize a different promoter site.

This binding process is almost like tying knots in an anchor rope. A diver could swim upward holding onto the rope, but a knot signals a spot to stop and decompress.

After the RNA polymerase arrives at a promoter site, a 17 base-pair segment of the double helix unwinds and the bases unpair. This unwinding converts a closed promoter complex to an open promoter complex. RNA polymerase is now ready to begin the RNA chain by incorporating the first nucleotide triphosphate. (Unlike DNA replication, no primer is necessary.) This first nucleotide triphosphate is usually a pppG or a pppA, which remains throughout transcription. This tag is at the new RNA molecule's 5' end, and growth begins when a new nucleotide links to the 3' position (see Figure 16-3).

After the first two nucleotides link (through the formation of the linking phosphate diester), the σ subunit leaves, allowing the core enzyme to bind more tightly to the substrate. A transcription bubble now forms that contains the RNA polymerase, the unwound portion of the DNA, and the rapidly forming nascent RNA. Initially, a short segment of the new RNA forms a hybrid helix with the DNA. This segment normally consists of about eight base pairs or one turn of the double helix. The growth rate is on the order of 50 nucleotides per second. (Compare this to DNA replication, which proceeds at about 800 nucleotides per second.)

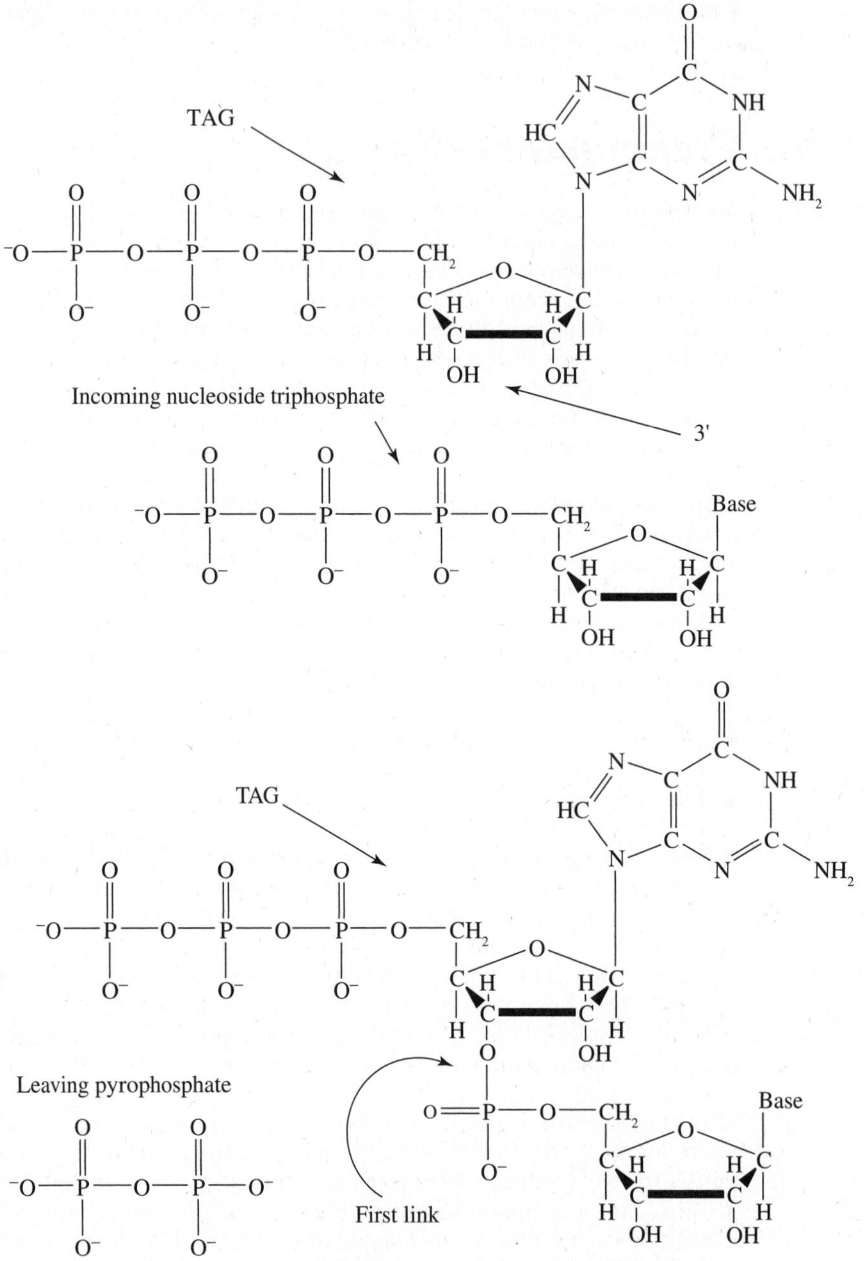

TAG

Incoming nucleoside triphosphate

3'

Base

Figure 16-3:
Linking of
the second
nucleotide
to the tag,
using pppG
as an exam-
ple (top),
and linked
nucleo-
tides at
the chain's
beginning
(bottom).

TAG

Leaving pyrophosphate

First link

Base

RNA polymerase doesn't "proofread" the new RNA. Thus, errors creep in at a
higher rate than in replication. However, because the products don't always
pass to the next generation, there are fewer mutations or lasting effects. In any
case, the next RNA strand to form lacks this defect and behaves correctly. One

bad RNA in several hundred or more copies of the same gene is likely to have a minimal influence on the cell. In addition, RNA isn't as structurally stable as DNA, and thus, it isn't as long-lived (which can be a good thing).

Elongation proceeds until the RNA polymerase encounters a termination signal, initiating a series of actions. At this point, formation of new phosphate diesters ceases, the RNA-DNA hybrid separates, the portion of the DNA chain that's still open rewinds, and the RNA polymerase separates from the DNA.

Termination signals differ. One simple one is a palindromic (reading the same forward or backward) GC-rich region followed by an AT-rich region. The palindromic region is self-complementary, and these bases hydrogen bond to form a hairpin loop. The AT-rich region results in a number of U_{RNA}-A_{DNA} pairs, which have the weakest hydrogen bond interactions of all types of pairs. The formation of this hairpin and the AT region destabilizes, and the RNA-DNA hybrid and the nascent RNA begin to leave (see Figure 16-4).

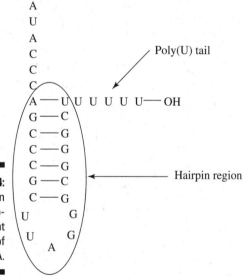

Figure 16-4: The hairpin and subsequent portion of the RNA.

Not all termination signals contain a hairpin and a U-rich segment. In at least some cases, RNA polymerase needs help. Evidence for this comes from the observation that *in vitro* RNA chains are often longer than *in vivo* chains for the same RNA. Clearly, the in vitro RNA polymerase is unable to terminate elongation. The missing aid is a protein known as the *rho factor* (ρ). This protein wraps about the nascent RNA soon after the RNA exits the transcription bubble. In the presence of RNA, the ρ protein hydrolyzes ATP, which supplies energy. The protein first attaches to an RNA segment that's poor in guanine and rich in cytosine. Rho moves along the nascent RNA until it encounters the transcription bubble. At this point, it breaks the RNA-DNA hybrid and separates the nascent RNA. Other proteins serve a similar function as the rho factor.

In prokaryotic cells, mRNA is either ready or nearly ready to function immediately after release from the transcription (translation may begin before transcription terminates). However, both tRNA and rRNA require cleavage and other modifications of the nascent RNA chain. Various nucleases cleave the RNA in a very precise manner. It's possible to get more than one protein from a long nascent RNA strand. Processing may require the connection of a number of nucleotides. For example, all tRNA molecules need a CCA tail to function correctly. Some cases may involve modification of the bases or ribose units.

Eukaryotic cells

Unlike in prokaryotic cells, transcription and translation occur in different regions of eukaryotic cells, leading to greater control of gene expression. Another difference is that eukaryotic cells extensively process mRNA in addition to rRNA and tRNA. After RNA polymerase action, mRNA acquires a cap and a poly(A) tail. Nearly all mRNA molecules are spliced. Splicing involves removal of introns with the remaining *exons* (segments that code for protein) being connected. Ninety percent of the nascent RNA may be introns.

Eukaryotic cells typically contain three types of RNA polymerase:

- ✔ Type I RNA polymerase, in the nucleolus, produces most forms of rRNA.

- ✔ Type II, in the nucleoplasm, produces mRNA and snRNA.

- ✔ Type III, in the nucleoplasm, produces tRNA and small rRNA molecules. (Actually, these polymerases only produce the pre-RNA forms of these molecules.)

Each of the three polymerases has a distinct type of promoter. These promoters may be in the same upstream sites as in prokaryotic cells, in downstream sites, or within the genes themselves. In addition to promoters, the polymerases may have enhancers. (We get a lot of e-mails trying to sell us enhancers.) Enhancers aren't promoters but they increase a promoter's effectiveness. Enhancers for a single promoter may occur in different positions on the DNA chain and are important for gene regulation. Both promoters and most enhancers are on the same side of the DNA chain as the gene they regulate; for this reason, they are *cis-acting* elements. The promoters, as discussed earlier, are typically a TATA box (usually between –30 and –100), CAAT box, and GC box (both are usually between –40 and –150). Enhancers may appear upstream, downstream, or within the gene about to undergo transcription. Enhancers that are present on the opposite DNA chain are *trans-acting* or *transcription* factors on the other DNA chain.

The typical series of events is that the transcription factor TFIID binds to the TATA box (TF stands for *transcription factor* and the II refers to *RNA polymerase II*). Binding is the result of a small component of TFIID known as *TBP* (TATA-box-binding protein), which has an extremely high affinity for the TATA box. When TBP binds to the DNA, substantial changes occur in the DNA, including some degree of unwinding.

Other components utilized in transcription later attach to the TBP. These are, in order: TFIIA, TFIIB, TFIIF, RNA polymerase II, and, finally, TFIIE. This final group is the *basal transcription complex.* This example illustrates only one of numerous transcription factor initiations.

In eukaryotic cells, nearly all, if not all, products of transcription (precursors) undergo further processing before they reach their final active form. In general, tRNA precursors have the 5' leader removed, splice to remove any and all introns, replace the poly(U) tail with a CCA sequence, and possibly modify some of the bases. Each of these processes requires one or more enzymes.

The precursors to the various forms of mRNA normally require the most modification. These precursors need, among other things, a 5' cap and a 3' poly(A) tail. The caps are cap 0, cap 1, and cap 2; the numbers refer to the number of methylated ribose sugars (see Figure 16-5). Caps aren't present on tRNA, snRNA, or rRNA.

Most mRNA has a poly(A) tail not encoded by DNA. Usually, addition of this tail is preceded by cleavage of an intron portion of the mRNA precursor. The series AAUAAA signals where the cleavage occurs. This series is only part of the signal; the other part is uncertain. After cleavage, a poly(A) polymerase adds about 250 adenylate residues to the 3' end. The exact purpose of the tail is uncertain, although it appears to enhance translation and increase the lifetime of the mRNA molecule.

In some cases, mRNA precursors need to be edited. *Editing* refers to an alteration of the base sequence other than that caused by splicing. An example is to chemically change one base into another. Editing occurs in the mRNA that encodes for apolipoprotein B (apo B). The entire protein contains 4,536 residues. However, a related 2,152-residue form is also important. The longer form, synthesized in the liver, is useful in the transport of lipids within the liver. The smaller form, synthesized in the small intestine, interacts with dietary fats. The same mRNA is responsible for both protein forms. In the small intestine, a deaminase acts on a specific cytosine and converts it to a uracil, which changes a CAA codon (Gln) to a UAA codon (stop), truncating the protein chain to yield the smaller form.

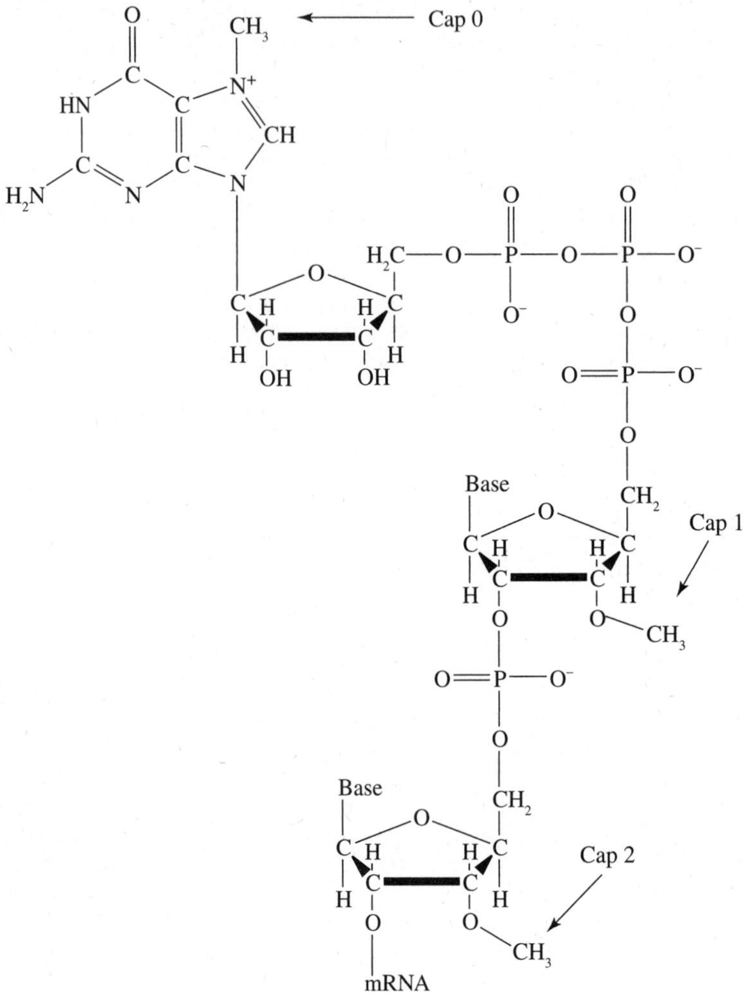

Figure 16-5:
The general
structure of
an mRNA
cap.

Splicing is a very common form of modification of all forms of RNA. *Splicing* involves the removal of introns and the joining of the exons to yield the final RNA molecule. Splicing must be very precise, as a miss by one base alters the entire sequence of codons present.

A number of different introns need to be removed. In eukaryotic cells, the intron begins with a GU and ends with an AG. Further refinement is present in vertebrates, where GU is the end of the sequence AGGUAAGU. A variety of AG sequences are found in higher eukaryotic cells. In general, one end of the intron loops about and connects to a point (the branch point) on the intron chain. Joining of the exons then proceeds.

Spliceosomes are important in the splicing of mRNA precursors. These assemblages contain the mRNA precursors, several snRNAs, and proteins known as *splicing factors.* A group of snRNAs labeled U1, U2, U4, U5, and U6 are important. U1 binds to the 5' end of the splice site and then to the 3' end. U2 binds to the branch point, U4 blocks U6 until the appropriate moment, U5 binds to the 5' splice site, and U6 catalyzes the splicing. Alternate splicing procedures also occur.

Alternate splicing can lead to production of different proteins from the same RNA.

Not a Secret Any Longer: The Genetic Code

Just as DNA serves as the template for the generation of RNA, mRNA serves as the template for the generation of protein. To synthesize the appropriate protein, a species needs to interact with this template to assure the incorporation of the correct amino acid. The interaction species is tRNA. This relatively small form of RNA has two important regions: a template recognition site and the appropriate amino acid. The template recognition site is an anticodon, which is complementary to a codon on the mRNA. Attachment of the amino acid to the tRNA is by the action of an aminoacyl-tRNA synthetase. Each of the 20 amino acids has at least one specific synthetase. This enzyme attaches the specific amino acid to the 3' terminal adenosine of the tRNA (see Figure 16-6).

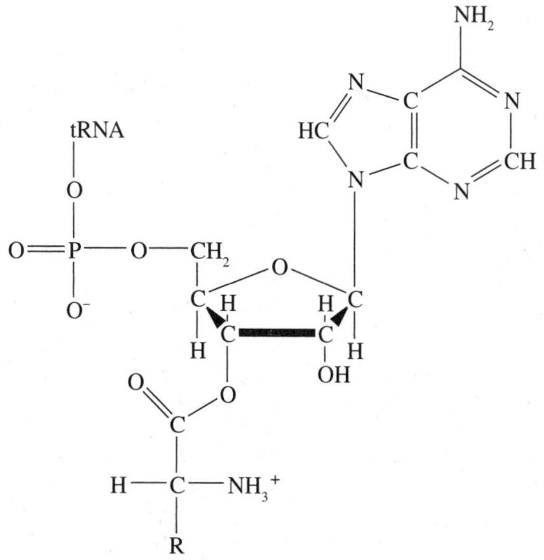

Figure 16-6:
The attachment of an amino acid to the terminal adenosine.

Codons

The genetic code contains the information necessary for the synthesis of proteins and consists of a set of three-letter words made from an alphabet containing four letters. Each three-letter word is a *codon*. This vocabulary is universal as it applies to all known living organisms.

The four letters are as follows:

- ✔ A, for adenine
- ✔ C, for cytosine
- ✔ G, for guanine
- ✔ U, for uracil

The four letters yield a dictionary containing a total of 64 words. Sixty-one of these words code for specific amino acids, and the remaining three words code for no amino acid. The codons that code for no amino acid are the *stop signals*.

Because there are only 20 amino acids to code for, the presence of 61 codons means that some amino acids can come from more than one codon.

Table 16-1 lists the genetic code.

Table 16-1			The Standard Genetic Code				
Codon	Amino Acid	Codon	Amino Acid	Codon	Amino Acid	Codon	Amino Acid
AUA	Ile	ACA	Thr	AAA	Lys	AGA	Arg
AUC	Ile	ACC	Thr	AAC	Asn	AGC	Ser
AUG	Met	ACG	Thr	AAG	Lys	AGG	Arg
AUU	Ile	ACU	Thr	AAU	Asn	AGU	Ser
CUA	Leu	CCA	Pro	CAA	Gln	CGA	Arg
CUC	Leu	CCC	Pro	CAC	His	CGC	Arg
CUG	Leu	CCG	Pro	CAG	Gln	CGG	Arg
CUU	Leu	CCU	Pro	CAU	His	CGU	Arg
GUA	Val	GCA	Ala	GAA	Glu	GGA	Gly
GUC	Val	GCC	Ala	GAC	Asp	GGC	Gly
GUG	Val	GCG	Ala	GAG	Glu	GGG	Gly

Codon	Amino Acid	Codon	Amino Acid	Codon	Amino Acid	Codon	Amino Acid
GUU	Val	GCU	Ala	GAU	Asp	GGU	Gly
UUA	Leu	UCA	Ser	UAA	*Stop*	UGA	*Stop*
UUC	Phe	UCC	Ser	UAC	Tyr	UGC	Cys
UUG	Leu	UCG	Ser	UAG	*Stop*	UGG	Trp
UUU	Phe	UCU	Ser	UAU	Tyr	UGU	Cys

Analysis of the genetic code shows that two amino acids — methionine (see Figure 16-7) and tryptophan — only have one codon each. At the other extreme, three amino acids — arginine, leucine, and serine — each has six codons. The remaining fifteen amino acids have at least two codons each. Amino acids with more codons are more abundant in proteins. Examining Table 16-1 shows that most *synonyms* (codons that code for the same amino acid) are grouped together and differ by a single base, usually the last base in the codon. The similarity of synonyms limits potential damage from mutations. Other correlations are present in the table; see what others you can find.

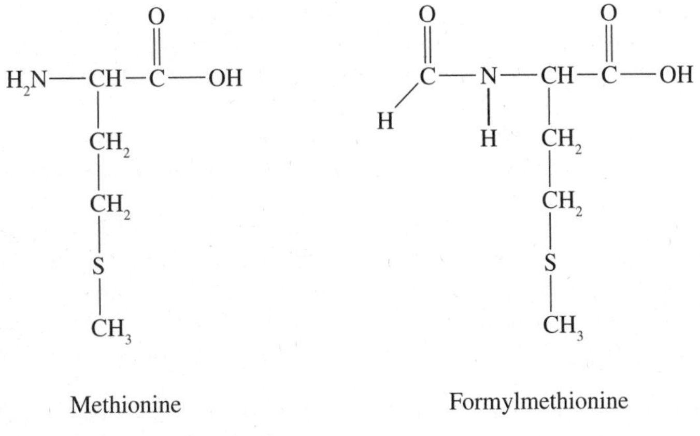

Figure 16-7: Structures of methionine and formylmethionine.

Methionine   Formylmethionine

Alpha and omega

Although tRNA doesn't read the termination sequences (UAA, UAG, and UGA), specific proteins known as *release factors* read them. When a release factor binds to the ribosome, it triggers the release of the new protein, and release of the protein signals new synthesis to begin.

The stop signals are rather obvious in Table 16-1, but what about the start? What signals the initiation of protein synthesis? The initiation sequence is usually AUG, the codon for methionine (see Figure 16-8). In eukaryotic cells, additional factors come into play. In many bacteria, fMet (formylmethionine) is the initial amino acid, which AUG usually codes for; however, GUG works sometimes.

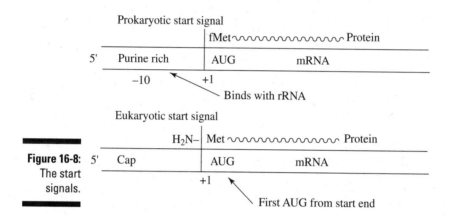

Figure 16-8: The start signals.

The genetic code is nearly universal; the codons correspond to the same amino acid in most cases. A few exceptions are known. For example, the code in mitochondrial DNA has several differences from nuclear DNA. In mitochondrial DNA, UGA is not a stop signal but a codon for tryptophan.

In prokaryotic cells, coding for proteins is continuous, but that's not always true in the case of eukaryotic cells. In some mammals and birds, most genes are discontinuous. For example, in the gene encoding for β-globin, some regions don't encode for a portion of protein. The gene contains about 1,660 base pairs. About 250 pairs on each end plus an additional 500 pair segments code for protein (exons). Two segments, one of about 120 base pairs and one of about 550 base pairs, don't code for protein (introns). The entire gene has, in sequence, a 240-pair exon, a 120-pair intron, a 500-pair exon, a 550-pair intron, and a 250-pair exon.

If an mRNA forms from a gene containing introns, it needs to undergo modification before it's useful. The intron regions from the mRNA must be cut and the exon ends must be spliced together to form the final mRNA molecule, which is then read without any problem by the tRNA. In most cases, the intron portion begins with a GU and ends with a pyrimidine-rich segment ending with an AG. This combination signals the intron domain.

Models of Gene Regulation

An organism doesn't need to produce all the different proteins all the time. To control which proteins form at which time requires some form of *gene regulation.* For example, it would be terribly inconvenient if women produced milk proteins all the time, instead of only when they're lactating after having given birth. When the organism requires a specific protein, it "switches on" a certain gene, and when a sufficient quantity of that protein is present, the gene must be "switched off." Control may occur either at the transcription level (gene regulation) or at the translation level.

In this section we examine processes in prokaryotic cells and then move on to the more complicated processes that take place in eukaryotic cells. The examination of the simpler mechanisms in prokaryotic cells gives insight into the processes in eukaryotic cells — the basic processes are similar.

As usual, our prokaryotic example is *E. coli.* Researchers gained insights into gene regulation when they changed the diet of *E. coli* from glucose-rich to lactose-rich. For the cells to utilize this alternate energy source, they must generate the enzyme β-galactosidase. This enzyme is normally available at very low levels, a situation that quickly changes after replacing the glucose with lactose. One clue to the mechanism was that as the levels of β-galactosidase increased, so did the levels of galactoside permease (which transports lactose into the cell) and thiogalactoside transacetylase (which detoxifies other materials transported by galactoside permease). Thus, one change in the environment triggered multiple enzymes. This coordinated triggering of gene expression is called an *operon.*

The Jacob-Monod (operon) model

The simultaneous change in the levels of three different enzymes by one change in the environment suggested a link between the control mechanisms, and thus, biologists Francois Jacob and Jacques Monod proposed the operon model to explain gene regulation. This model requires a regulator gene that affects a number of structural genes and an operator site. The operator and associated structural genes constitute the operon. The regulator gene is responsible for producing a *repressor protein.* The repressor protein binds to the operator site and prevents expression of the structural genes, as shown in Figure 16-9. (The *etc.* in the figure simply means that there may be more structural genes that follow the same pattern.)

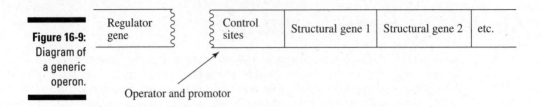

Figure 16-9:
Diagram of
a generic
operon.

The multiple structural genes produce one large mRNA, and this single RNA strand is capable of generating a set of proteins. An mRNA that's capable of encoding for multiple proteins is *polygenic* or *polycistronic*.

The lac operon

One of the better understood operons, the *lac operon* is the model regulatory system that, since its discovery in 1961, has provided extensive insight into how a cell regulates its genome. Figure 16-10 illustrates the lac operon.

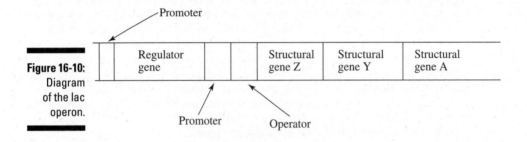

Figure 16-10:
Diagram
of the lac
operon.

The *lac operator* is a palindromic DNA sequence with a twofold symmetry axis. The repeat isn't always a perfect palindrome. (Many protein-DNA interactions involve a matching of symmetry.) The lac operator is as follows, with the center axis in bold:

5'-TGTGTGGAATTGTGAGC**GG**ATAACAATTTCACACA-3'

3'-ACACACCTTAACACTCG**CC**TAATGTTAAAGTGTGT-5'

The *lac repressor* is a dimeric protein that can join to form a tetramer. In the absence of lactose, the repressor tightly binds to the operator. The presence of the repressor prevents RNA polymerase from unwinding the DNA strand to initiate transcription — in essence, keeping the transcription turned off.

The presence of lactose isn't the direct trigger of the lac operon; the trigger is *allolactose*. Both lactose and allolactose are disaccharides composed of galactose and glucose (see Figure 16-11). Lactose has an α-1,4 linkage, whereas in allolactose the linkage is an α-1,6. Allolactose results when the few molecules of β-galactosidase that are normally present in the cell first encounter lactose. This disaccharide, along with a few similar molecules, is an inducer of the lac operon. The inducer binds to the repressor and reduces the affinity of the latter to the operator on the DNA. With its affinity reduced, the repressor detaches from the operator, and the DNA segment is now open for business.

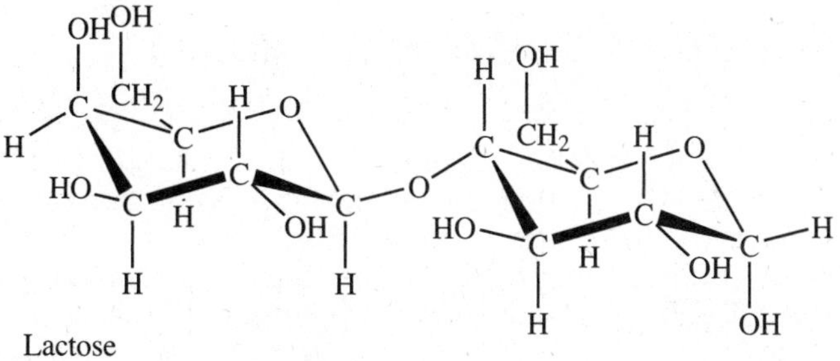

Lactose

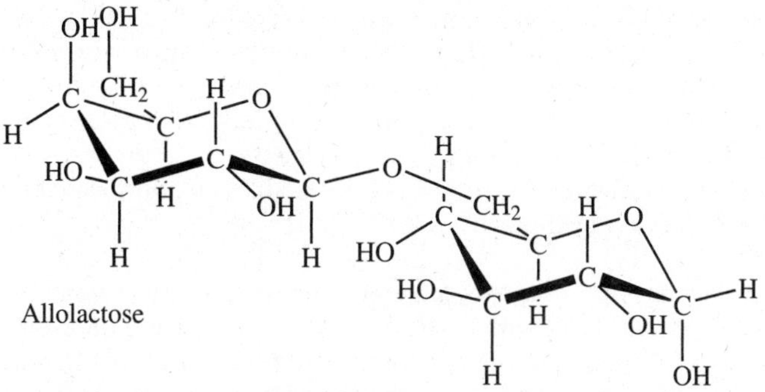

Figure 16-11: Lactose and allolactose are disaccharides.

Allolactose

When transcription begins, all three structural genes become active, and the cell begins producing β-galactosidase, galactoside permease, and thiogalactoside transacetylase. This continues until the lactose and allolactose concentrations fall so that the repressor proteins are available to reattach to the DNA.

Other prokaryotic regulators

The *pur repressor* affects the genes responsible for the biosynthesis of purines and, to a lesser extent, pyrimidines. This protein is similar in structure to the lac repressor; however, the pur repressor only binds to the operator after another molecule binds to the repressor. Therefore, while the binding of another molecule releases the lac repressor, the binding of another molecule causes the pur repressor to bind. The other molecule has an opposite effect. In the case of the pur repressor, the other molecule is a *co-repressor*.

Other regulators stimulate transcription instead of repressing it. The *catabolite activator protein* (CAP) is one example. This protein interacts with the promoter and, along with two cAMP molecules, interacts with RNA polymerase. This interaction leads to stimulating the initiation of transcription of certain genes.

Regulation of eukaryotic genes

The regulation of genes in eukaryotic cells is similar to but more complex than regulation in prokaryotic cells. One reason for this is that the typical eukaryotic genome is much larger than the typical prokaryotic genome. Another source of complexity is that many eukaryotic cells are part of a larger organism and don't serve the same purpose that other cells do within the same organism. For example, although some of the proteins are the same, a liver cell must produce a different overall set of proteins than a heart cell produces.

Histones

Eukaryotic DNA has a group of proteins associated with it. These small, basic proteins are called *histones.* They're basic because approximately 25 percent of the amino acid residues present are either arginine or lysine. These are tightly bound to the DNA and total approximately half of the mass of a chromosome. A complex of the cell's DNA and associated protein is *chromatin,* which has five important histones: H1, H2A, H2B, H3, and H4 (the latter four associate with one another).

Chromatin apparently consists of repeat units that have two copies each of H2A, H2B, H3, and H4, with a strand of DNA consisting of about 200 base pairs tightly wrapped around this histone octamer. Each of these repeating units is a nucleosome. The wrapping of the DNA to form a nucleosome yields a significant compaction of the DNA. Research indicates that about 145 of the 200 base pairs are actually associated with the histone octamer, and the remaining base pairs are linker DNA that link one histone octamer to the next. Histone H1 usually binds to linker DNA.

The eight histones in a histone octamer are arranged into a tetramer with the composition $(H3)_2(H4)_2$ and two dimers each with the composition (H2A)(H2B). All the histone proteins have long tails rich in arginine and lysine residues that extend out of the core. Modification of these tails is important for gene regulation.

The structure of chromatin is a factor in eukaryotic gene regulation. For a gene to be available for transcription, the tightly packed chromatin structure must open so that transcription machinery can get in and do its job. In addition, the structure regulates access to regulatory sites on DNA. Enhancers disturb this structure, which explains why enhancers can have an effect on the expression of a gene even though the enhancer site may be thousands of base pairs away from the gene. Certain enhancers only occur in specific types of cells. Thus, the genes they enhance are only expressed in these cells. For example, the gene to produce insulin is expressed only in pancreatic cells.

A modification of DNA can also inhibit gene expression. Approximately 70 percent of the 5'-CpG-3' sequences in mammals have the cytosine methylated. The distribution of the methylated cytosines (see Figure 16-12) varies with cell type. Regions in chromatins necessary for gene expression in that cell are *hypomethylated* (have fewer methylated cytosines) relative to similar regions in cells where no expression of the gene occurs. The presence of the methyl group interferes with the binding of enhancers and promoters.

All cells contain the same genetic info, but their use of that genetic info differs. A neuron does what neurons do, and a liver cell does what liver cells do. Thus, the genetic info must be regulated differently in all cell types. In addition, even similar cell types may need to access genetic information for protein production at different times of life — during fetal development, puberty, and so on — and this genetic info, even within the same cell types, is accessed differently. Enhancers and methylation help access genetic info differently both temporally and spatially throughout the multicellular body.

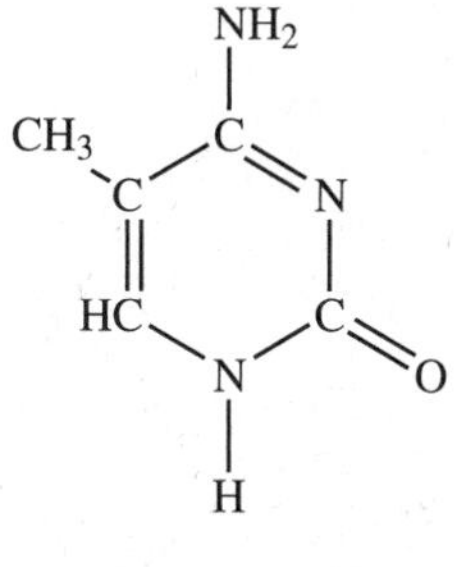

Figure 16-12:
Structure of methylated cytosine.

Mediating transcription

Eukaryotic cells require a variety of transcription factors to initiate transcription; no factor can carry out the entire process on its own. This cooperation of factors is called *combinatorial control,* and it's necessary in organisms with multiple cell types and is helpful in other eukaryotic cells.

A number of nonpolar molecules, such as the steroid hormones, can easily pass through the hydrophobic cell membrane and bind to receptor proteins. They're very specific. Estrogen (see Figure 16-13) is one example of a steroid hormone. Such molecules are known as *ligands.*

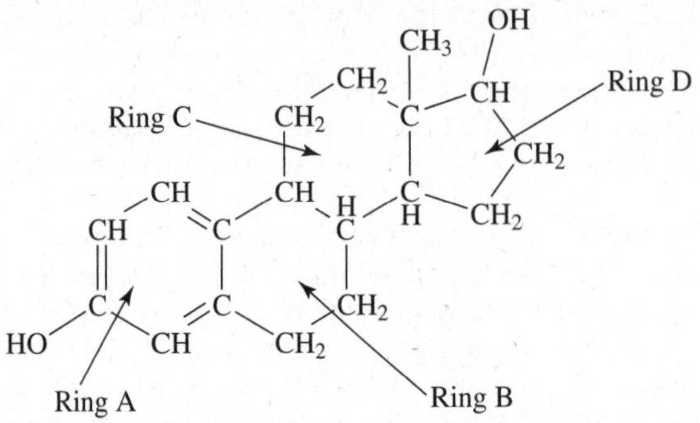

Figure 16-13:
Structure of
estrogen.

The ligand binds to a specific site — called, helpfully, the *ligand-binding site* — that's present near the end of a receptor protein. This portion of the protein contains many nonpolar residues that have an affinity for hydrophobic molecules. Most steroids bind to nuclear hormone receptors. In these cases, a DNA binding site near the protein's center contains eight cysteine residues that are necessary to bind zinc ions, four residues for each. The zinc ions stabilize structure and give the area the name *zinc finger domains* (other cysteine residues and zinc ions are nearby). The binding of a molecule to the ligand-binding site causes a significant structural rearrangement of the protein. This situation would seem to be similar to the lac repressor in prokaryotic cells; however, experiments indicate that there's no significant alteration in binding affinity.

The next part of the puzzle involves a number of small proteins known as *coactivators*. Near the center of each of these are three regions with the pattern Leu-X-X-Leu-Leu. Each of these regions generates a short hydrophobic α-helix. These three helixes bind to a hydrophobic region on the ligand-binding region. The presence of the ligand appears to enhance the binding of a coactivator. (A receptor protein may act as a repressor, especially in the presence of a co-repressor.)

Just what are the roles of coactivators and co-repressors? Their effectiveness appears to be linked to their ability to covalently bond to the tails of the histones. Histone acetyltransferases (HATs) catalyze this modification of the histone tails (a process that's reversed by histone deacetylase enzymes — see Figure 16-14). This process changes a very polar (positively charged lysine) amide to a much less polar (neutral) amide, resulting in a significant reduction in the affinity of the tail to the associated DNA. To a lesser degree, this process reduces the affinity of the entire histone to the associated DNA, which allows access of a portion of the DNA to transcription.

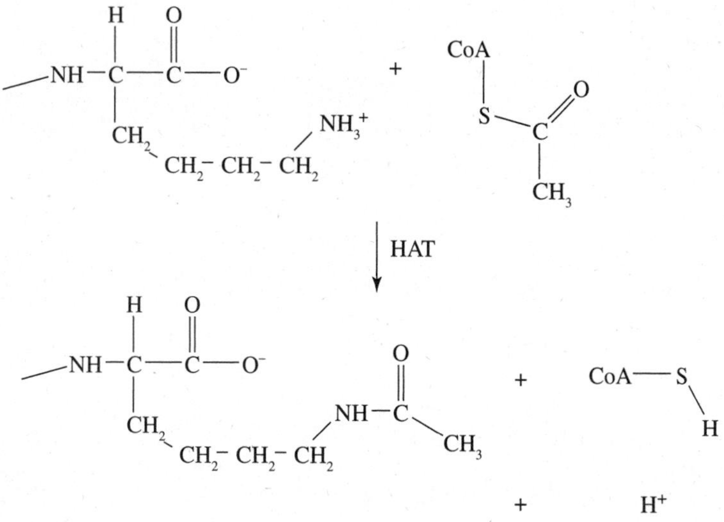

Figure 16-14:
Reaction
catalyzed
by histone
acetyltrans-
ferases
(HATs).

The acetylated lysine residues also affect the acetyllysine-binding domain
(the *bromodomain*) present in many of the eukaryotic transcription
regulatory proteins.

There are two important bromodomain-containing proteins. One of these is
a large complex of ten proteins that binds to the TATA-box-binding protein
that's responsible for the transcription of many genes. The other proteins
containing bromodomains are part of large complexes known as *chromatin-
remodeling engines*. As the name implies, these proteins alter the structure
of the chromatins, which changes their behavior.

All these factors alter the availability of portions of the DNA structure to transcription. After the DNA becomes open, the procedures we discuss earlier in
the chapter come into play.

Chapter 17

Translation: Protein Synthesis

You're no doubt familiar with the process of *translation* — converting text from one language into another. The process of translation in biochemistry does exactly the same thing. In this chapter, we explain translation and its place in the synthesis of proteins.

Hopefully Not Lost in Translation

In biochemical *translation,* the four-letter alphabet of the nucleic acids becomes the twenty-letter alphabet of proteins. In doing so, genetic information is passed on. Translation occurs in the cell's *ribosomes,* which contain *ribosomal RNA* (rRNA). The information necessary for translation travels from the cell nucleus to the ribosomes via *messenger RNA* (mRNA). The messenger RNA binds to the smaller ribosome, and transfer RNA (tRNA) brings amino acids to the mRNA.

Why translation is necessary

The purpose of translation is to put together specific amino acids in a specific order to produce a specific protein (so specific!). Messenger RNA provides the template or blueprint for this process. To utilize this template, something must bring the amino acids to the mRNA, and that something is transfer RNA.

Transfer RNA has two important sites. One site is for the attachment of a specific amino acid. For example, only one specific type of tRNA transfers the amino acid methionine. The other site is the recognition site, which contains an *anticodon* — a sequence of three bases that are complementary to a codon on the mRNA. A codon sequence of AUG on the mRNA matches the UAC anticodon on a tRNA. All this takes place in the ribosome, home of rRNA.

Home, home in the ribosome

The *ribosome* is the factory that produces proteins. Thousands of ribosomes are present in even the simplest of cells. They are complex units composed of RNA and protein. It's possible to dissociate a prokaryotic ribosome into two units. One unit is the 50S, or large unit, and the other is the 30S, or small unit. The large unit contains 34 different proteins, labeled L1 through L34, and two RNA molecules, labeled 23S and 5S. The small unit contains 21 different proteins, labeled S1 through S21, and an RNA molecule labeled 16S.

A prokaryotic ribosome contains three rRNA molecules (23S, 16S, and 5S), one copy of proteins S1 through S21, two copies of L7 and L12, and one each of the other L1 through L34 proteins. L7 and L12 are identical except that L7 has an acetylated amino terminus. S20 and L26 are identical. Mixing the constituents in vitro leads to the two subunits reconstituting themselves. A version of the structure of the 16S form of ribosomal RNA appears in Figure 17-1.

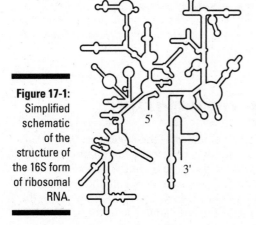

Figure 17-1:
Simplified
schematic
of the
structure of
the 16S form
of ribosomal
RNA.

The Translation Team

A number of players, along with the rRNA, must interact to form a protein molecule. In addition, the ribosome's structure is important to controlling protein synthesis. Both the rRNA and the protein molecules control this structure. One possibly helpful analogy is the game of football.

The team captain: rRNA

RNA makes up approximately two-thirds of a ribosome's mass. The three rRNA units play a key role in the ribosome's shape and function (the proteins apparently fine-tune the ribosome's shape and structure). The three rRNAs form from the cleaving and processing of transcribed 30S RNA. A significant portion of each of the rRNA molecules has numerous *duplex regions* (short stretches of base-paired RNA).

The 30S and 50S rRNA subunits combine to form a 70S ribosome, which holds an mRNA in place during translation. The ribosome also has three sites for various tRNA molecules: the E, P, and A sites.

- The E site is the *exit site*. A tRNA occupies this position after delivery of its amino acid and just before exiting the ribosome.

- The P site is the *peptidyl site,* which holds the tRNA that contains either the initial amino acid or the C-terminal amino acid of a protein chain.

- Finally, the A site is the *aminoacyl site,* which holds the tRNA attached to the next amino acid in sequence.

When the 30S and 50S subunits join, they create A and E sites at the interface of the subunits. The P site of the 50S unit is the opening of a tunnel through which the growing protein chain passes out of the ribosome.

Here's the snap: mRNA

The base sequence of the mRNA is read in the 5' → 3' direction, and transcription occurs in this same direction. (Prokaryotic cells sometimes take advantage of this by beginning translation before transcription is over. This situation can't occur in eukaryotic cells because transcription and translation are physically separated.) The mRNA resulting from transcription gains a cap and a poly(A) tail before it ventures out of the nucleus on its trip to the ribosome. (See Chapter 16 for more on transcription.)

Translation doesn't begin at the 5' terminus of the mRNA molecule. Just as there's a "stop" signal to terminate translation, there's a "start" signal. The 5' terminus base-pairs with the 3' terminus of the 16S rRNA. This region is normally about 30 nucleotides in length (a portion of this region, called the *Shine-Dalgarno sequence,* is purine-rich).

Shortly after this sequence is the start signal. In most cases, the start signal is AUG (methionine), though in some instances the signal is GUG (valine). In *E. coli,* the first amino acid is formylmethionine instead of methionine. The formylmethionine is usually removed soon after translation begins. Prokaryotic cells may have more than one start and stop signal because many of the mRNA molecules are *polygenic* (or *polycistronic*) — that is, they produce more than one protein. Figure 17-2 illustrates the structures of methionine and formylmethionine attached to tRNA.

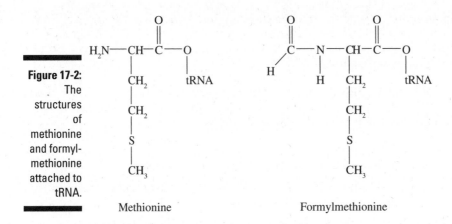

Figure 17-2:
The
structures
of
methionine
and formyl-
methionine
attached to
tRNA.

Methionine   Formylmethionine

Carrying the ball: tRNA

Several features are common to all forms of tRNA. Each form of tRNA is a single strand that contains between 73 and 93 nucleotides. tRNA has between seven and fifteen unusual bases (not one of the usual four — A, C, G, or U) in each molecule. Approximately half of the nucleotides present are base-paired. The activated amino acid is attached to the hydroxyl group at the 3'-end of the chain. The hydroxyl group is on the adenosine residue of a CCA segment. The other end, the 5'-end, is phosphorylated. The phosphorylation usually is a pG (phosphorylated G). The anticodon is contained in a loop near the molecule's center. The molecule somewhat resembles a cloverleaf in shape.

Many of the unusual bases are methylated or dimethylated forms of A, C, G, or U that are usually the result of post-transcriptional modification of the molecule. The presence of the methyl groups interferes with the formation of some base pairs, which prevents certain additional interactions. Methyl groups are nonpolar, so their presence makes regions of the tRNA hydrophobic, which affects their interaction with ribosomal proteins and syntheses. The unusual bases include dihydrouridine, dimethylguanosine, inosine, methylguanosine, methylinosine, pseudouridine, and ribothymidine. Inosine, shown in Figure 17-3, is part of the anticodon. Many of these unusual bases are in or near the bends in the structure of tRNA.

Five regions are not base-paired; these are shown in Figure 17-4. (Note that the structure of tRNA shown in Figure 17-4 isn't the actual three-dimensional structure of tRNA.) Starting at the 5'-end, the unpaired regions are, in order, the DHU loop, the anticodon loop, the extra arm, the TψC loop, and the 3-CCA terminus. (The name of the DHU loop derives from the presence of several dihydrouracil residues. The anticodon loop contains the segment that recognizes the codon on the mRNA, and the extra arm contains a variable number of residues. The TψC loop derives its name from the presence of the sequence ribothymine-pseudouracil-cytosine.) These loops make each tRNA different, even though the overall structure is the same. The loop also allows for its function — one end interacts with the mRNA and the other end interacts with the amino acid.

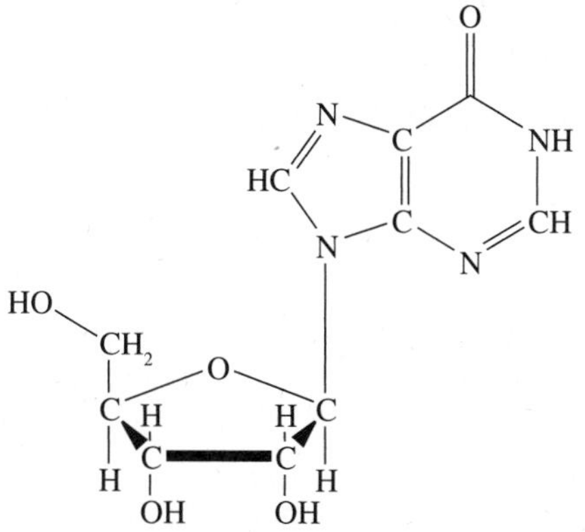

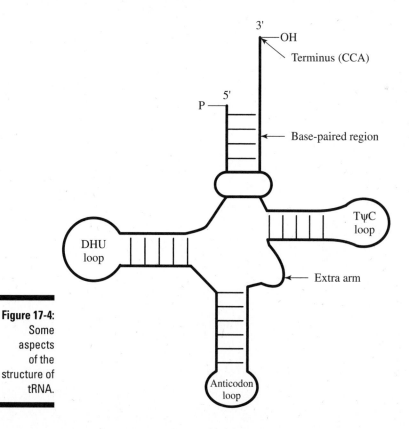

Figure 17-3:
The structure of inosine.

Figure 17-4:
Some aspects of the structure of tRNA.

The anticodon is present in the 5' → 3' direction, and it base-pairs to a codon in the 3' → 5' direction. This matches the first base of the anticodon with the third base of the codon. (Don't forget the convention of writing base sequences in the 5' → 3' direction.)

Charging up the middle: Amino acid activation

It's imperative that the correct amino acid attaches to the tRNA because the presence of an incorrect amino acid or the absence of any amino acid is devastating to translation. Connection of the amino acid to the tRNA activates the amino acid. Joining free amino acids is a nonspontaneous process; however, connecting the amino acid to the tRNA changes the free amino acid to a more reactive amino acid ester. The amino acid-tRNA combination is an *aminoacyl-tRNA* or a *charged tRNA* (see Figure 17-5).

Specific aminoacyl-tRNA synthetases, called *activating enzymes,* catalyze the activation reaction. The process begins with an amino acid and an ATP forming an aminoacyl adenylate (see Figure 17-6), which leads to the release of a pyrophosphate.

Each amino acid has a separate aminoacyl-tRNA synthetase.

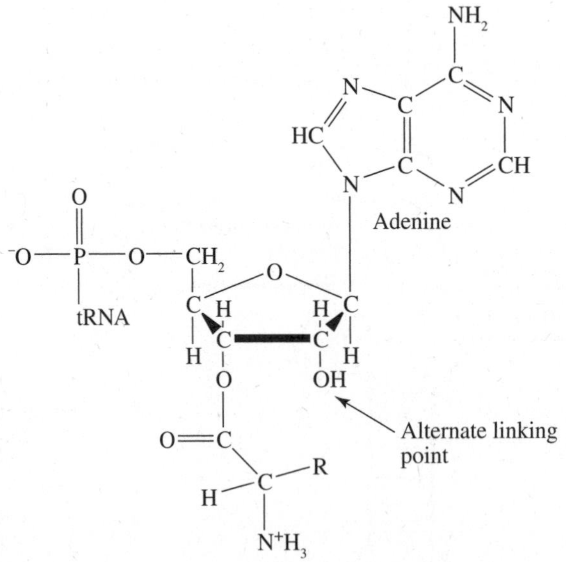

Figure 17-5: An example of an aminoacyl-tRNA.

Adenine

Alternate linking point

tRNA

NH_2

N^+H_3

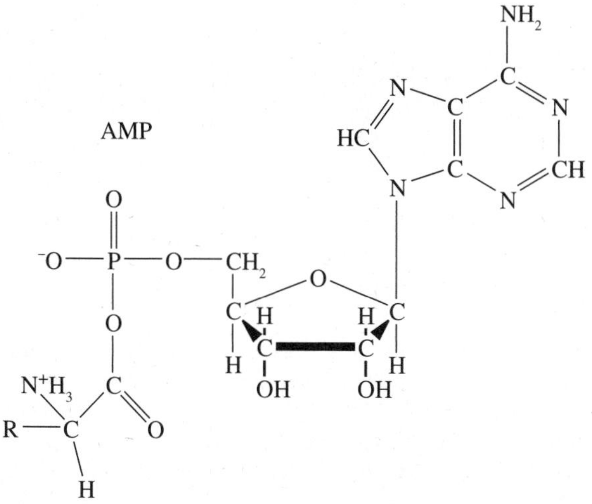

Figure 17-6:
Structure
of an
aminoacyl
adenylate.

The two classes of aminoacyl-tRNA synthetases are denoted Class I (monomeric) and Class II (usually dimeric). Each class is responsible for ten amino acids. The CCA arm adopts different structures when it interacts with members of the different classes, and ATP adopts a different conformation when it interacts with members of different classes. Most Class II examples attach the amino acid as Figure 17-5 illustrates, whereas Class I examples attach the amino acid to the alternate linking site. (Refer to Figure 17-4 to see some aspects of the structure of tRNA.)

The conversion of an aminoacyl adenylate, once formed, remains tightly bound to the synthetase until it can form an aminoacyl-tRNA.

To make sure that the aminoacyl-tRNA synthetase incorporates the correct amino acid, the enzyme must take advantage of specific properties of the amino acids. Examining the amino acids serine, valine, and threonine can give some insight into the selection process. These three amino acids appear in Figure 17-7, where they're drawn to emphasize similarities in the side chain. (Recall that the threonine side chain is chiral but the others are not.) These amino acids have size differences (–H for –CH$_3$) and hydrogen bonding differences (–OH can, but –CH$_3$ can't). The recognition site has the proper size and composition to take advantage of these specific properties. A significant species in this site is a zinc ion, which coordinates to the enzyme and the amino acid.

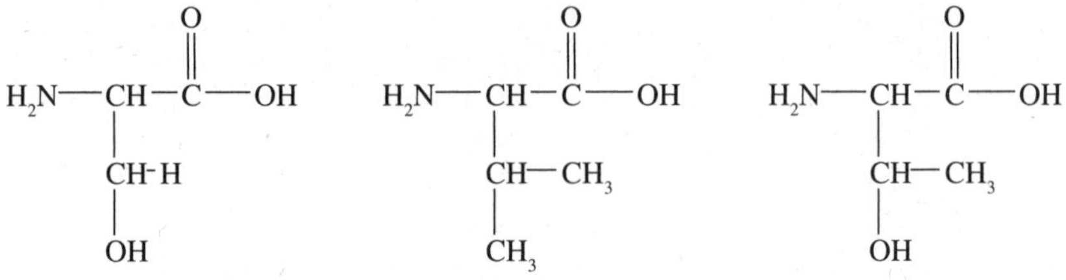

$$
\begin{array}{ccc}
\overset{\displaystyle O}{\underset{\displaystyle \parallel}{}} & \overset{\displaystyle O}{\underset{\displaystyle \parallel}{}} & \overset{\displaystyle O}{\underset{\displaystyle \parallel}{}} \\
H_2N\!-\!CH\!-\!C\!-\!OH & H_2N\!-\!CH\!-\!C\!-\!OH & H_2N\!-\!CH\!-\!C\!-\!OH \\
\mid & \mid & \mid \\
CH\text{-}H & CH\!-\!CH_3 & CH\!-\!CH_3 \\
\mid & \mid & \mid \\
OH & CH_3 & OH \\
Serine & Valine & Threonine
\end{array}
$$

Figure 17-7:
Structures of serine, valine, and threonine.

Even with these differences, serine sometimes replaces threonine. Fortunately, the enzyme includes an editing feature. The editing site is near the reaction site, but it's not the same. Similar editing occurs in other aminoacyl-tRNA synthetases. Amino acids, such as tryptophan, don't have closely similar analogues; thus, editing is far less important in these cases.

The aminoacyl-tRNA synthetases need to be able to recognize the anticodon present to make sure they interact with the appropriate tRNA, matching it to the correct amino acid. The enzymes may recognize other features of the tRNA structure such as the size of the extra arm and the hydrophobic character imparted by methylating some of the ribonucleotides.

Hooking Up: Protein Synthesis

The major steps in protein (polypeptide) synthesis are as follows:

1. **Activation**

2. **Initiation**

3. **Elongation**

4. **Termination**

These basics apply to all living organisms; no differences exist between human translation, fungi translation, and tulip translation. Synthesis proceeds from the amino to the carboxyl direction of the protein.

In this section we discuss these synthesis steps in greater detail. These steps involve tRNA, mRNA, and rRNA, along with a number of protein factors.

Activation

As we mention earlier in this chapter, during *activation*, an amino acid reacts with ATP to give aminoacyl adenylate. The aminoacyl adenylate then reacts with a specific tRNA to give aminoacyl-tRNA plus AMP. This constitutes one of the players necessary for the translation game.

Initiation

During *initiation,* an mRNA attaches to a ribosome by interacting, through the Shine-Dalgarno sequence, with the 30S rRNA subunit. Then the anticodon of the first tRNA attaches to the AUG (or GUG) codon on the mRNA. This codon occupies the P site of the 30S subunit. The amino acid extends into the P site of the 50S subunit. The 30S and 50S portions of the rRNA combine to produce the 70S ribosome (clamping down, basically, allowing a structural space for the mRNA, tRNA, and AAs to interact appropriately). The combination of the two subunits allows the tRNA to interact with both parts.

To initiate translation, the mRNA and the first tRNA must be brought to the ribosome. Three proteins, known as *initiation factors,* accomplish this task: IF1, IF2, and IF3. First, the 30S ribosome subunit, IF1, and IF3 form a complex. The two initiation factors bound to the 30S subunit interfere with a premature joining of the 30S and 50S subunits without the necessary mRNA. The remaining initiation factor, IF2, binds to GTP. The IF2-GTP combination binds to the initiator-tRNA, and the IF2-GTP-initiator-tRNA unit binds to the mRNA. Interaction of the Shine-Dalgarno sequence and the 16S rRNA manipulates the incoming group into the correct position.

Combining all these units with the 30S subunit gives the 30S initiation complex. Hydrolysis of GTP as the 50S subunit approaches leads to expulsion of the initiation factors. With the initiation factors out of the way, the remaining moieties join to give the 70S initiation complex. (Wow, trying saying that three times fast!) After this complex forms, elongation can begin. (I bet you hope that your brain elongates to hold all this new information!)

Elongation

During *elongation,* a second activated tRNA comes into the A site (which is adjacent to the P site) on the 30S subunit, where it binds to the appropriate codon. A protein known as *elongation factor Tu* or *EF-Tu* brings the activated tRNA to the A site. EF-Tu forms a complex with the activated tRNA (in the GTP form), and this complex protects the ester linkage holding the amino acid to the tRNA. In addition, the complex doesn't allow the activated tRNA to enter the A site if there's no codon-anticodon match. EF-Tu interacts with all tRNAs except the initiator-tRNA. The energy needed for the EF-Tu to leave the tRNA in the ribosome comes from the hydrolysis of the GTP unit induced by the protein known as *elongation factor Ts.*

The two amino acids extend into the ribosome's peptidyl transferase center. The amino group of the aminoacyl-tRNA from the A site is held in position to attack the ester linkage of the aminoacyl-tRNA in the P site. The catalyzed formation of the peptide bond occurs, accompanied by separation from the tRNA in the P site. The protein is now attached to the A site (30S).

With the loss of its amino acid, the tRNA no longer interacts in the same way with the ribosome. The tRNA moves to the E site of the 50S subunit as the next RNA, with its attached polypeptide, moves to the P (tunnel) site of the same subunit. The ribosome must now move over (the fancy way to say it is *translocate*) by one codon. For translocation to occur, the elongation factor G enzyme is needed. (*EF-G* or *translocase* is the protein that aids translocation.) The hydrolysis of GTP to GDP supplies the energy for the move. This move places the polypeptide-tRNA into the P site of the 30S subunit. At the same time, the amino acid–stripped tRNA disengages from the mRNA and moves into the E site of the same subunit. Throughout this process, the polypeptide chain remains in the P site of the 50S subunit.

The first tRNA leaves the E site. Now the elongation cycle can begin again with the entry of another tRNA carrying the next amino acid. The process continuously cycles until a codon signals "stop."

Termination

A "stop" signals *termination,* which results in the release of the protein, the last tRNA, and the mRNA.

Recall that the stop signals are UAA, UGA, and UAG.

Normal cells don't contain tRNAs with anticodons complementary to these codons. However, proteins known as *release factors* (RF) recognize these three codons. Release factor 1, RF1, recognizes UAA and UAG. Release factor 2, RF2, recognizes UAA and UGA. Release factor 3, RF3, is an intermediary between RF1, RF2, and the ribosome. The release factors carry a water molecule into the ribosome in place of an amino acid. The final reaction, the one that releases the newly formed protein, is the hydrolysis of the last ester linkage to a tRNA. The water brought in by the release factors is necessary for this hydrolysis.

The 70S ribosome remains together for a short time. Dissociation of the complex is mediated by a ribosome release factor and EF-G. GTP supplies the energy for this process.

The wobble hypothesis

Experimental studies have found that even pure tRNA molecules are capable of recognizing more than one codon. Biochemists developed the *wobble hypothesis* to explain this behavior, and subsequent work has firmly established this hypothesis.

The presence of the unusual base, inosine (refer to Figure 17-3), in the anticodon loop is the key to understanding the wobble hypothesis. This base is capable of base-pairing with adenine, cytosine, or uracil, allowing for some variation, or wobble, in the matching of codon to anticodon. The presence of inosine increases the number of different codons a particular tRNA can read. The first two bases in the codon pair to the corresponding bases in the anticodon. The third base is the wobble position.

Review the table of codons (Table 16-1 in Chapter 16) and see which amino acids depend only on the first two bases. *Hint:* Look at valine.

Table 17-1 shows the base-pairing rules for the wobble hypothesis. The presence of an A or C as the first base allows the reading of only one codon. The presence of a G or U allows the reading of two codons, and an I allows the reading of three codons. Inosine is a useful base for allowing wobble; however, as Table 17-1 shows, no wobble occurs only when the first anticodon base is an A or a C. In general, the base in the wobble position forms weaker hydrogen bonds than normal because of the strain in the environment. The weaker hydrogen bonding aids in the loss of the tRNA after it delivers its amino acid.

Table 17-1   Base-pairing Rules for the Wobble Hypothesis

Base on Anticodon (First Base)	Bases Recognized on Codon (Third Base)
A	U
C	G
G	U, C
U	A, G
I	U, C, A

Four codons code for valine, comprising a four-codon family. If you examine three of the codons for valine — GUU, GUC, and GUA — they all pair to the anticodon CAI instead of the anticodons CAA, CAG, and CAU. For this reason, one CAI anticodon replaces three other anticodons. The remaining valine codon is GUG, which requires the synthesis of only two types of tRNA instead of four. Other four-codon families also work this way.

The only cases where the codons for a particular amino acid differ in the first two bases are the six-codon families: arginine, leucine, and serine. These families require three different tRNAs.

The presence of wobble reduces the number of necessary tRNAs in a cell from 61 to 31. However, cells usually have some number of tRNAs between these extremes. All the tRNAs coding for a specific amino acid require only one aminoacyl-tRNA synthetase.

Variation in Eukaryotic Cells

All cells follow the same basic pattern for translation. However, eukaryotic cells show some variations. More proteins are necessary to mediate translation, and the steps are, in general, more complicated.

Ribosomes

In eukaryotic cells, the ribosomes contain a 60S subunit and a 40S subunit that combine to produce an 80S ribosome. The 40S subunit contains an 18S rRNA analogous to the 16S in the 30S subunit. The 60S subunit has three rRNA components: a 5S and a 23S, analogous to the 5S and the 23S of the prokaryotic 50S subunit, and a unique 5.8S rRNA.

The Human Genome Project

The U.S. Human Genome Project was begun in 1990. It was originally scheduled to last for 15 years, but because of rapid advances in the field of biotechnology, it finished two years ahead of schedule in 2003. The U.S. Department of Energy and the National Institutes of Health coordinated the project.

Goals

The project had the following goals:

- Identify all the 20,000 to 25,000 genes in human DNA.

- Determine the sequences of the approximately 3 billion base pairs in human DNA.

- Store the information in databases.

- Improve data analysis tools.

- Transfer the developed technology to the private sector.

- Address the ethical, legal, and social issues associated with the project.

In addition to human DNA, researchers also studied the genetic blueprints of *E. coli*, a common bacterium found in humans as well as mice and fruit flies. The goal of transferring the technology to the private sector was included to develop the infant biotechnology industry and to encourage the development of new medical applications.

Potential Benefits

Some potential benefits of the Human Genome Project include the following:

- Improved disease diagnosis

- Earlier detection of genetic predispositions to disease

- Drug design and gene therapy

- Creation of new biofuels

- More effective ways of detecting environmental pollutants

- ✔ Studying evolution through mutations in lineages

- ✔ Forensic identification of subjects through DNA analysis

- ✔ Establishing paternity

- ✔ Matching organ donors and patients

- ✔ Creation of insect- and disease-resistant crops

- ✔ Creation of biopesticides

- ✔ Increased productivity of crops and farm animals

Many of these benefits show up in people's lives every day.

Ethical, Legal, and Social Issues

One of the unique aspects of the Human Genome Project was that it was the first large scientific project that studied and addressed potential ethical, legal, and social implications that arose from the data generated by the study. Questions such as the following were addressed:

- ✔ Who should have access to personal genetic information?

- ✔ Who controls and owns genetic information?

- ✔ How reliable and useful is fetal genetic testing?

- ✔ How will genetic tests be checked for reliability and accuracy?

- ✔ Do parents have the right to test their children for adult-onset diseases?

- ✔ Do people's genes influence their behavior?

- ✔ Where is the line between medical treatment and enhancement?

- ✔ Are genetically modified foods safe for humans?

Many questions have been raised, but as yet, few answers have resulted.

Initiator tRNA

In eukaryotic cells, the initiator amino acid is methionine instead of formyl-methionine. As in prokaryotic cells, a special tRNA is necessary for the first tRNA — a modification of the normal methionine-carrying tRNA.

Initiation

AUG is the only initiator codon in eukaryotic cells, and this is always the AUG nearest the 5' end of the mRNA. There's no purine-rich sequence immediately before this as in prokaryotic cells. The 40S ribosome subunit attaches to the mRNA cap and moves base by base in the 3' direction until it reaches an AUG codon. The hydrolysis of ATP by helicases powers this process. Many more initiation factors are present in eukaryotic cells. A eukaryotic initiation factor has the symbol eIF instead of IF.

Elongation and termination

The EF-Tu and EF-Ts prokaryotic elongation factors have the eukaryotic counterparts EF1α and EF1βγ. Translocation is driven by eukaryotic EF2 with the aid of GTP. Only one release factor, eRF1, is present in eukaryotic cells, unlike the two factors in prokaryotic cells. To prevent the reassembly of the two ribosome subunits, eIF3 functions like the IF3 protein in prokaryotic cells.

Part VI
The Part of Tens

The 5th Wave   By Rich Tennant

"Yeah, those are the Carboxylic brothers, and you know
what their alcohol absorption rate is."

In this part . . .

We wrap things up by zooming out a bit and looking at topics we haven't covered yet. Here we compile two chapters' worth of short and sweet information about some of the lesser-known potential applications of biochemistry and some perhaps unexpected careers related to it.

Chapter 18

Ten Great Applications of Biochemistry

*I*n this chapter, we briefly look at some of the biochemical applications and tests that have changed people's everyday lives. Although these examples are just a few of the hundreds we could have chosen, we feel that all these have made and continue to make a significant impact on society. And we hope you realize that scientists are discovering more applications almost daily.

Ames Test

The *Ames test* determines whether a substance affects (mutates) the structure of DNA. In this test, salmonella bacteria is exposed to the chemical in question (food additives, for example), and scientists measure changes in the way the bacteria grows. Many substances that cause mutations in this bacteria also cause cancer in animals and humans, so this test is used today to screen chemicals for their potential to cause cancer.

Pregnancy Testing

Pregnancy tests come in two types: One uses a urine sample and the other uses a blood sample. Both detect the presence of the hormone *human chorionic gonadotropin* (hCG). The placenta produces this hormone shortly after implantation of the embryo into the uterine walls. The hormone accumulates rapidly in the body in the first few days after implantation. Home pregnancy urine tests are typically around 97 percent accurate if done two weeks after implantation. Blood tests, performed in a clinic, are more costly but can detect pregnancy as early as a week after implantation.

HIV Testing

Tests have been developed to screen for the presence of the *human immunodeficiency virus* (HIV), the virus that causes *acquired immune deficiency syndrome* (AIDS). These tests may be done on urine, blood serum, or saliva, and they detect HIV antigens, antibodies, or nucleic acids (RNA). The nucleic acid tests (NAT) detect a 142-base sequence located on one of the HIV genes. Most blood banks use a combination of tests to ensure accuracy.

Breast Cancer Testing

Most breast cancer isn't hereditary, but in 5 to 10 percent of cases, there *is* a heredity linkage. The vast majority of these cases involve mutations in two genes: the breast cancer-1 gene (BRCA1) and the breast cancer-2 gene (BRCA2), which were discovered in 1994 and 1995, respectively. Females who inherit a mutation in either one of these genes have a greatly increased chance of developing breast cancer over their lifetime. Positive tests for these mutations allow the individual to schedule screening tests at a more frequent rate than the general population.

Prenatal Genetic Testing

In *prenatal genetic testing,* tests are commonly performed on blood or tissue samples from the fetus to screen for potential genetic defects. These tests may involve *amniocentesis* — collection of a sample of amniotic fluid that contains cells from the fetus — or collection of blood from the umbilical cord. Tests such as these are used to detect chromosomal abnormalities such as Down syndrome or birth defects such as spina bifida.

PKU Screening

Phenylketonuria (PKU) is a metabolic disorder in which an individual is missing an enzyme called *phenylalanine hydroxylase.* Absence of this enzyme allows the buildup of phenylalanine, which can lead to mental retardation. All states in the United States require PKU testing of all newborns. Infants who test positive are placed on a diet low in phenylalanine, allowing them to develop normally. Check out cans of soft drinks and you'll find a warning on many of them that they contain phenylalanine.

Genetically Modified Foods

Biochemists have developed the ability to transfer genes from one organism into other organisms, including plants and animals. This allows the creation of crops that are more pest- and disease-resistant and animals that are more disease-resistant. Genetic modification can also be used to increase the yield of milk, eggs, or meat. In 1993, the U.S. Food and Drug Administration issued a license for the first genetically modified food for human consumption. It was a new tomato called Flavr Savr, which was resistant to rotting. However, the public has been slow to accept genetically altered foods because of concerns about unforeseen effects on the population and environment.

Genetic Engineering

Genetic engineering involves taking a gene from one organism and placing it into another. The recipient may be a bacterium, a plant, or an animal. One of the most well-known examples of genetic engineering involves the hormone insulin. Diabetes used to be treated with insulin derived from pigs or cows, and although they're very similar to human insulin, these animal-derived insulins aren't identical, and they caused problems for some individuals. Biochemists solved this problem by inserting the gene for human insulin into bacteria. The bacteria, through the process of *translation,* created human insulin. (See Chapter 17 for much more on translation.)

Cloning

In 1996, Dolly the sheep was cloned — the first mammal ever cloned from adult animal cells. The cloned sheep was, of course, genetically identical to the original adult sheep. Scientists created this clone by taking cells from the udder of a 6-year-old ewe and growing them in the lab. They then took unfertilized eggs and stripped the genetic material from them. Finally, they inserted the genetic material of the other 6-year-old ewe they had grown in the lab into these stripped cells and implanted them into the uterus of another sheep. And voilà! — Dolly was born. Since Dolly, scientists have successfully cloned many other animals, but the idea of cloning a human has caused worldwide debate that will surely continue for decades.

Gene-Replacement Therapy

In *gene-replacement therapy* (or often more broadly called *gene therapy*), a modified or healthy gene is inserted into an organism to replace a disease-causing gene. A virus that has been altered to carry human DNA is often used to deliver the healthy gene to the patient's targeted cells. This process was first used successfully in 1990 on a 4-year-old patient who lacked an immune system because of a rare genetic disease called *severe combined immunode-ficiency* (SCID). Individuals with SCID are prone to life-threatening infections. They lead isolated lives, avoiding people and commonly taking massive doses of antibiotics. Scientists removed white blood cells from the patient, grew them in the lab, and inserted the missing gene into the cells. They then inserted this genetically altered blood back into the patient. The process allowed the child to develop normally and even attend school, but the treatment must be repeated every few months.

Chapter 19

Ten Biochemistry Careers

*B*ecause of recent advances in biochemistry and its related field, biotechnology, many new professions are available for students majoring in biochemistry. Those who stop at the BS degree often find themselves working as technicians in a variety of industries, but for those who go on to earn their MS or PhD, many more opportunities are possible.

Graduates at all levels find positions in a wide variety of career areas, including forensics, industrial chemistry, molecular biology, pharmacology, technical sales, virology, horticulture, immunology, forestry, and so on. We mention several careers throughout the book, so here we include careers one may not normally associate with the field of biochemistry.

Research Assistant

Research assistants work in the area of biochemical research and development as part of a team. They may investigate new genetic tests, be involved in genetic engineering or cloning, or help with the development of new drugs or drug protocols. In addition to performing typical technical biochemical procedures, research assistants analyze data and prepare technical reports and summaries. Research assistants are often also involved in the search for inventions that can lead to patents. They may eventually head up their own research groups.

Plant Breeder

Plant breeders design and implement plant-breeding projects in conjunction with other research teams. They may be involved in the development of disease-resistant strains of crops, or they may search for ways to increase crop yields using biochemical and biotechnological techniques. They may also be involved in personnel management, public relations, and/or advising their company about future projects and plant-breeding goals.

Quality Control Analyst

Quality control analysts conduct analyses of raw materials and the finished products coming off the production line. They collect data concerning quality control test procedures and pinpoint sources of error. Along with quality control engineers, analysts ensure that the quality of a product remains high. As you may imagine, maintaining high quality is especially important when the product is a genetically modified virus or a genetically altered food.

Clinical Research Associate

Clinical research associates design and implement clinical research projects such as a new drug protocol or the use of a new virus for gene therapy. They may travel to the various field sites where the clinical trials are being conducted to coordinate and/or supervise the trials. The clinical associate analyzes and evaluates data from the trials to ensure that clinical protocols are followed. A background in nursing or pharmacology is useful.

Technical Writer

Anyone who has ever read a poorly written set of directions or technical manual realizes the importance of a good technical writer. Technical writers in the biochemical world write and edit operating procedures, lab manuals, clinical protocols, and other materials. They ensure that these documents are written in a way that meets government regulations. They may develop professional development programs for staff members and write news releases. Part of their job is to take highly technical reports and edit them in such a way that they're understandable to the company's administration and the general public.

Biochemical Development Engineer

Biochemical development engineers take the biochemical process developed in the laboratory and scale it up through the pilot plant stage to the full production plant. They help determine what instrumentation and equipment are needed and troubleshoot problems in the scale-up procedure. They work to develop more efficient manufacturing processes while maintaining a high degree of quality control. They may also be involved in technological advances from another area and apply them to their manufacturing process.

Market Research Analyst

Market research analysts analyze and research a company's market, its variety of products, and its competition. They perform literature searches and make presentations on technical areas and new potential markets for the company. They predict future marketing trends based on market research and may even be involved in the preparation of research proposals.

Patent Attorney

Patent attorneys coordinate and prepare documentation for patent applications. They track a company's research studies and recommend the timing of patent filings. They collect supporting documentation and negotiate patent licenses and other legal agreements. They may become involved with interference and appeal hearings.

Pharmaceutical Sales

An individual with a degree in biochemistry is a natural for a career in pharmaceutical sales. These sales representatives spend much of their time on the road, talking to hospital personnel, physicians, pharmacists, and others. They're quite familiar with their company's products and try to be as persuasive as possible in touting their advantages over the competition. They have to be familiar with statistics and issues of concern in the medical community in order to successfully communicate with potential clients.

Biostatistician

Biostatisticians are statisticians who work in health-related fields. They design research studies and collect and analyze data on problems, such as how a disease progresses, how safe a new treatment or medication is, or what the impact is of certain risk factors associated with medical conditions. They may also design and analyze studies to determine healthcare costs and healthcare quality. They're instrumental in the design stages of studies, providing expertise on experimental design, sample sizes, and other considerations.

Index

• B •

• C •

• F •

• G •

• *H* •

• O •

• P •

Apple & Macs

iPad For Dummies
978-0-470-58027-1

iPhone For Dummies,
4th Edition
978-0-470-87870-5

MacBook For Dummies, 3rd
Edition
978-0-470-76918-8

Mac OS X Snow Leopard For
Dummies
978-0-470-43543-4

Business

Bookkeeping For Dummies
978-0-7645-9848-7

Job Interviews
For Dummies,
3rd Edition
978-0-470-17748-8

Resumes For Dummies,
5th Edition
978-0-470-08037-5

Starting an
Online Business
For Dummies,
6th Edition
978-0-470-60210-2

Stock Investing
For Dummies,
3rd Edition
978-0-470-40114-9

Successful
Time Management
For Dummies
978-0-470-29034-7

Computer Hardware

BlackBerry
For Dummies,
4th Edition
978-0-470-60700-8

Computers For Seniors
For Dummies,
2nd Edition
978-0-470-53483-0

PCs For Dummies,
Windows
7th Edition
978-0-470-46542-4

Laptops For Dummies,
4th Edition
978-0-470-57829-2

Cooking & Entertaining

Cooking Basics
For Dummies,
3rd Edition
978-0-7645-7206-7

Wine For Dummies,
4th Edition
978-0-470-04579-4

Diet & Nutrition

Dieting For Dummies,
2nd Edition
978-0-7645-4149-0

Nutrition For Dummies,
4th Edition
978-0-471-79868-2

Weight Training
For Dummies,
3rd Edition
978-0-471-76845-6

Digital Photography

Digital SLR Cameras &
Photography For Dummies,
3rd Edition
978-0-470-46606-3

Photoshop Elements 8
For Dummies
978-0-470-52967-6

Gardening

Gardening Basics
For Dummies
978-0-470-03749-2

Organic Gardening
For Dummies,
2nd Edition
978-0-470-43067-5

Green/Sustainable

Raising Chickens
For Dummies
978-0-470-46544-8

Green Cleaning
For Dummies
978-0-470-39106-8

Health

Diabetes For Dummies,
3rd Edition
978-0-470-27086-8

Food Allergies
For Dummies
978-0-470-09584-3

Living Gluten-Free
For Dummies,
2nd Edition
978-0-470-58589-4

Hobbies/General

Chess For Dummies,
2nd Edition
978-0-7645-8404-6

Drawing
Cartoons & Comics
For Dummies
978-0-470-42683-8

Knitting For Dummies,
2nd Edition
978-0-470-28747-7

Organizing
For Dummies
978-0-7645-5300-4

Su Doku For Dummies
978-0-470-01892-7

Home Improvement

Home Maintenance
For Dummies,
2nd Edition
978-0-470-43063-7

Home Theater
For Dummies,
3rd Edition
978-0-470-41189-6

Living the
Country Lifestyle
All-in-One
For Dummies
978-0-470-43061-3

Solar Power Your Home
For Dummies,
2nd Edition
978-0-470-59678-4

Available wherever books are sold. For more information or to order direct: U.S. customers visit www.dummies.com or call 1-877-762-2974.
U.K. customers visit www.wileyeurope.com or call (0) 1243 843291. Canadian customers visit www.wiley.ca or call 1-800-567-4797.